AF566684

ENVIRONMENT MANAGEMENT AND AUDIT

Edited by

P. SASI BHUSHANA RAO
Senior Reader, P.G. Department of Life Sciences,
S.K.C.G. College, Parlakhemundi,
Gajapati, Orissa.

P. MOHANA RAO
Reader, P.G. Department of Commerce,
S.K.C.G. College, Parlakhemundi,
Gajapati, Orissa

Foreword by

Prof. P.K. SEN-SARMA
Emeritus Scientist, Calcutta

DEEP & DEEP PUBLICATIONS PVT. LTD.
F-159, Rajouri Garden, New Delhi-110 027

ISBN 81-7629-248-6

Typeset by LASER TECH, A-31A, Rajouri Garden,
New Delhi-110027 • Phone : 5930798, 5469995

Printed in India by ELEGANT PRINTERS, A 38/2, Phase I,
Mayapuri, New Delhi-110064

Published by DEEP & DEEP PUBLICATIONS PVT. LTD.,
F-159, Rajouri Garden, New Delhi-110027 • Phone: 5435369, 5440916

Contents

Foreword

Environment is the sum total of all conditions that are essentially required for the survival of life forms. Thus, environmental processes form the basis of resource management and economic development. Absence of environmental consideration tends to cause serious ecological damage. Further, precise knowledge of various facets of sound environmental management is the *sine qua non* for sustainable development which has been succinctly defined by the Brundtland Commission (1987) as "the meeting of needs of the present without compromising the ability of the future generations to meet their own need". Poverty, lack of resources, population pressure and global inequity of the resource use are creating unparalleled social and environmental problems at national and global levels. Thus, sustainable development tends to strike a balance between the demands of economic development and the need for management of environmental processes, be it waste recycling, renewal of life sustaining elements like water, air and soil. These processes are obviously inter-connected and inter-dependent. It is ironic that significance of changes militating against the environmental processes is often not realized by the power that be till the damage becomes almost irreversible.

Some of the parameters that are posing formidable challenges for the sustainable development operated through appropriate linkages of economic and ecological processes are: Population pressure of human kind and the domesticated herbivores, unabated deforestation, shortage of grass and grazing lands, loss of biodiversity through deforestation and other activities immical to ecology and environment, shortage of water due to unplanned water management including harvesting of underground water, poor soil management, improper solid waste management, unplanned mining leading to soil damage, urban trap, climatic changes, changes in cultural dimensions, habitat destruction for wild-life resulting in growing lists of extinct and endangered species of plants and animals.

Global human population level has already touched six billion. The United Nations have projected that population will range between 7.2 to 11.2 billion heads in the year 2050, the mid-level having been pegged at 9.2 billion. Demographic growth of such a dimension would create enormous pressure on the environmental resource base and the eco-system. It is estimated that chronic water shortage will affect almost three billion people globally in 52 countries including India. Shortage of clean fresh water in India is likely to be acute in view of the increased pollution level in our fresh water sources due to contamination arising from untreated waste water, industrial and agrochemicals, deforestation, soil erosion, etc. There is, however, an increasing realisation on the importance of water for sustainable development and to find ways and means to manage this precious resource more effectively and wisely. The picture in the Indian forestry sector is also gloomy. Though recorded forest cover is 76.52 Mha, actual forest cover currently does not exceed 63.3 Mha, which is merely, 19.3% of the total land area out of which only 11.2% has forests having a crown density higher than 40%. Pristine and untouched forests are practically non-existing. The causes for this dismal picture seem to be large scale illicit felling, overgrazing, encroachments, unsustainable management practices, including inadequate facilities to keep down recurrent forest fires etc. India is one of the twelve mega-biodiversity centres in the world with 46,000 plant and 80,000 known animal species and it contributes 8% of known global biodiversity with only 2.4% of the world's land area. But the biodiversity is under increasing threat from habitat loss, reckless over exploitation of genetic resources, increasing illicit trade in wild life products, and poaching. Already 1,500 plant species and 141 animal species (79 mammals, 44 birds, 15 reptiles and 3 amphibians) and several species of invertebrates are endangered. It is needless to state that the conservation of biodiversity is important from ethical, bio-ecological and economic development aspects. Though the Convention on Biological Diversity (C.B.D.) decided in Rio in 1992, (Rio Earth Summit) that the developed nations should contribute 0.77% of their GNP for environment clean-up programs in poor nations, it is hardly observed by the developed nations which are the greatest polluters of the environment. In the Indian industrial sector, the picture is far from satisfactory, as only ninety-four industrial firms have environment management system (EMS) conforming to ISO 1400 standard and another thirty-five claims to get it soon. Adherence to ISO 1400 would ensure safer and cleaner production contributing to green ecology. Ironically, personnel handling EMS are civil engineers mostly in addition to their own duties and professionals are hardly hired. The concept of environment audit, though somewhat nebulous in India, is an important aspect. Section 14 of the Environment (Protection) Rules 1986 as amended from time to time mandates submission of environment audit reports, but in practice the submissions do not fulfil the requirements of the pollution control laws that *inter alia* include a statement including data on pollution generated/release of pollutants, quantities of hazardous wastes generated, characteristics of waste material, impact of pollution control measures undertaken and proposals for additional investment for environmental protection and abatement of pollution. Notwithstanding these, correct definition for environmental audit should also include social and welfare aspects in addition.

The book, Environment Management and Audit, encompasses aspects as discussed in the preceding paragraphs through twenty-nine chapters. Each chapter has been written by a subject matter specialist. It is gratifying to observe that through the chapters, authors have been able to identify the critical question of environmental management and auditing including environmental laws and their enforcement. Inclusion of some of the important aspects such as mineral development and environment, environment friendly sustainable agriculture, industrial environment and its dimensions, environment education, management of water pollution, genetic diversity and its conservation, environment auditing, energy audit, social and environmental accounting and auditing and the role of NGOs in environmental amelioration is topical and deserves commendation.

The editors, Drs. P. Sasi Bhushana Rao and P. Mohana Rao, have vast experience on the topic and have done a commendable job in bringing together a vast array of specialists so as to present a multi-disciplinary approach to this complex and emerging discipline of science. It is hoped that the contents of the book will be able to dispel the false notion that environmental management aspects are immical to developmental process, as the emphasis in the book is on sustainable development. Environmental and socal accounting and auditing are inescapable aspects of augering social welfare and cultural equity in addition to efficient management of the environment.

The book is a welcome addition to the growing publications on the environment, ecology, conservation of genetic diversity etc.

Therefore, this book will undoubtedly by useful to the farmers, foresters, educationists, lawyers, judges of the green bench, planners, administrators, industrialists, professionals, politicians, accountants, auditors, researchers, students and lay public.

PROFESSOR P.K. SEN-SARMA
F.N.A.Sc, F.A.Sc, F.A.A.S., F.A.S.T.,
F.I.A.W.S., F.E.S.I., F.M.A.
Emeritus Scientist, Calcutta.

Preface

The words ecology (house study) and economics (house management) are both derived from the same greek word *Oikos* meaning house. Till recently ecologists and economists deliberately looked away from each other resulting in today to an ecological crisis. Had the economist and ecologist worked together in the past, our environment which may be defined as sum total of conditions that affect the life and development of organisms would not have been what it is today and is not the same what it was earlier. Industrial revolution is a blatant disregard to the code of environmental ethics. The finite natural resource base of our environment is already impigned by our activities aimed at economic growth, with little concern for the quality of the environment. Presently we are at the cross roads of a serious dilemma—The Economic growth, Sustainable development and Environment for the next millennium. The already degraded environment calls for its diligent management. Environmental audit is one prerequsite for an efficient management to pave the way for sustainable development and environment.

It is our endeavour to give this book a multi-disciplinary approach. This is a compendium of articles on science, social science and commerce. Articles on Bio-diversity, Resource Ecology, Environmental Education, Legislation, Tribal Economy, Management, Role of N.G.O.s and Auditing are contributed by academicians and social workers from across the country. We hope this book will be a successful venture as a single forum elucidating varied dimensions of the subject, Environment Management and Audit to the reader.

We thank all those who helped in bringing out this book. We unfeigningly thank Deep and Deep Publications Pvt. Ltd. for their unsparing efforts in publishing this book. We sincerely appreciate the encouragement rendered by our family members.

P. SASI BHUSHANA RAO
P. MOHANA RAO

List of Contributors

Albert A. Vondra, CPA, is a Partner of Price Waterhouse's National Office in New York City. (Care: Journal of Accountancy).

A.C. Kulshreshtha, Central Statistical Organisation, Department of Statistics, New Delhi. (Care: The Management Accountant, ICWAI, 12 Sudder Street, Calcutta).

A. Jagannatha Raju, Managing Director, CCD, Parlakhemundi, District Gajapati, Orissa.

Amit Mitra, Free Lance Journalist and Research Consultant. (Care: Yojana, New Delhi).

B. Bharatha Lakshmi, Assistant Professor, Department of Zoology, Andhra University, Visakhapatnam, Andhra Pradesh.

B. Bhattacharyya, Dean, Indian Institute of Foreign Trade, New Delhi. (Care: Yojana, New Delhi).

C.P. Oberai, Inspector General of Forests, Ministry of Environment and Forests. (Care: Yojana, New Delhi).

Dilip Biswas, Chairman, Central Pollution Control Board, New Delhi, (Care: Yojana, New Delhi).

D.P. Satapathy, Department of Botany, Kharasrota College, Singhpur, Orissa.

Floyd A. Beams, (Care: The Journal of Accountancy, U.S.A.).

Gulab Singh, Central Statistical Organisation, Department of Statistics, New Delhi, (Care: The Management Accountant, ICWAI, 12 Sudder Street, Calcutta).

John P. Surma, CPA, Partner of Price Waterhouse in Pittsburgh, U.S.A. (Care: Journal of Accountancy, March 1992, pp. 51-55).

Jagabandhu Samal, Reader in Economics, Department of Economics, D.A.V. College, Koraput, Orissa.

Kodeeawara Ramanathan, School of Energy, Bharathidasan University, Tiruchirappalli.

Krishan Mahajan, Law Editor of Indian Express and the Advocate Commissioner of Supreme Court in the Ganga Pollution, Taj, Delhi Ridge, Leprosy Patients and Preservation of the Ghalif Zauq Tower Matters, (Care : Yojana, New Delhi).

Mangalendu Narayan Roy, Senior Lecturer, Department of Commerce, Berhampore College, Berhampore, Murshidabad, West Bengal.

Meera Singh, Faculty of Commerce, 129/6 Malviya Road, George Town, Allahabad.

M. Selvam, Centre for Distance Education, Bharathidasan University, Tiruchirappalli.

M.S. Swaminathan, Chairman, M.S. Swaminathan Research Foundation, Chennai (Care: Yojana, New Delhi).

M.V.G. Ahobala Rao, B.V.Sc., PGDBA, F. En. R.A. Int., 11-10-2 Chenchupet, Tenali.

Naba Kumar Patnaik, National Awardee (NCERT) Karana Street, Parlakhemundi, Gajapati, Orissa.

N.L.N.S. Prasad, Joint Director, Ministry of Environment and Forests, Eastern Regional Office, 194, Kharaval Nagar, Bhubaneswar, Orissa.

Nagesh Kini, Chartered Accountant based at Bombay, (Care: The Chartered Accountant, ICAI, Post Box 7100, Indraprasta Marg, New Delhi).

Paul E. Fertig, (Care: The Journal of Accountancy).

P. Khanna, Director, National Environmental Engineering Research Institute, Nagpur (Care: Yojana, New Delhi).

P. Ram Babu, Dy. Director, National Environmental Engineering Research Institute, Nagpur, (Care: Yojana, New Delhi).

P. Sasi Bhushana Rao, Senior Reader in Zoology, P.G. Department of Life Sciences, S.K.C.G. College, Parlakhemundi, Dist. Gajapati, Orissa.

R.K. Pachauri, Director, Tata Energy Research Institute (TERI), New Delhi, (Care: Yojana, New Delhi), TERI, Darbari Seth Block, Habitat Place, Lodi Road, New Delhi.

Rakesh Kumar, Department of Commerce and Business Administration, C/o Mr. P.N. Singh, B.H. II/216, T.B. Sapru Hospital, Allahabad, U.P.

P.K. Shetty, Environmental Studies Unit, National Institute of Advanced Studies, Indian Institute of Science Campus, Bangalore.

S.C. Lahiry, Deputy Advisor, Planning Commission, New Delhi, (Care: Yojana, New Delhi).

S.P. Adhikary, P.G. Department of Botany, Utkal University, Bhubaneswar, Orissa.

Trilok Kumar Jain, Faculty of Institute of Management Studies, Parakh Niwas, Veterinary Hospital Road, Bikaner, Rajasthan.

Uma Shanker Singh, Former Dy. Adviser, Planning Commission, (Care: Yojana, New Delhi).

Umesh Holani, Reader in Commerce, School of Studies in Commerce and Management, Jiwaji University, Gwalior, M.P.

The Editor, The Chartered Accountant, The Institute of Chartered Accountants of India, Post Box 7100, Indraprasta Marg, New Delhi.

The Editor, The Management Accountant, I.C.W.A.I., 12 Sudder Street, Calcutta.

The Editor, Yojana, Yojana Bhavan, Sansad Marg, New Delhi.

The Permissions Editor, The Journal of Accountancy, American Institute of Certified Public Accountants (AICPA), Harbour Side Financial Centre, 201 Plaza Three, Jery City, NJ-07311-3881 (201) 938 3000, U.S.A.

1

Environment and Trade: Some Issues

B. BHATTACHARYYA

With economic development comes the problems of environmental degradation. After decades of obsession with higher and higher rates of economic growth, the global society has now come to realise that what should be sought is sustainable development. However, sustainable development can be achieved only when all or at least majority of countries cooperate with each other. There are also potential trade-offs and conflicts which are difficult to reconcile. For example, preservation of rain forests is a matter of grave concern to the entire mankind as it is a key environmental resource. But for a country like Malaysia, which has the rain forest reserves, it is also an economic resource which can be used for the country's economic resource development. The country, therefore, is being asked to sacrifice its economic resource for global benefit. Such difficult issues are coming up in making the world progress towards sustainable development.

A large number of international agreements on environment have been negotiated which have substantial trade effects. Further, some

---

This article first appeared in Yojana, volume 41: No. 8, August 1997. Reprinted with permission, August 1997.

countries are taking unilateral steps directed towards environmental protection which also have trade implications for the partner countries.

The World Trade Organisation (WTO), GATT in its earlier incarnation, has also been involved in the post Uruguay Round period in studying where trade and environmental issues come together. This paper takes a look at some of these developments.

Many international conventions for preservation of enviornment and bio-diversity have been negotiated during the last three decades. Some of the important ones in which India is a party are discussed below:

CITES

Convention on International Trade in Endangered Species of Wild Fauna and Flora (CITES) was adopted on 3 March 1973 and entered into force on 1 July 1975. It has established a global system of controls on cross-border transactions in threatened animals, plants and specimens derived from them. CITES has used trade measures to protect species threatened with extinction and species which may become endangered unless trade is strictly regulated. Trade in these species must be authorised by the designated authorities in its signatory countries. Under the export/import policy, India prohibits export of species of CITES. Similarly, import of wild animals including their parts and products as well as ivory are banned in India. It is, however, extremely difficult to stop poaching and illegal trade because the enforcement infrastructure currently available is less than what is required. Still the regulatory agencies have made some headway in reducing illegal hunting and trade in endangered animals and plants.

BASEL CONVENTION

Basel Convention on the control of transboundary movements of hazardous wastes and their disposal was adopted on 22 March 1989 and came into force on 5 May 1992. The convention seeks to address the major problems with respect to the disposal of hazardous and toxic wastes, such as transboundary movements of such wastes, including disposal at sea and shipment from developed to developing countries. At the 2nd meeting of the convention held in March 1994, it was decided to immediately prohibit transboundary movements of hazardous wastes which is destined for final disposal from OECD to non-OECD countries. Another resolution was adopted to phase out transboundary movements of hazardous wastes for recycling or recovery purpose by 31 December 1997.

Some Indian industries have been importing such wastes for recycling or other industrial uses for quite some time. Such imports have now been prohibited by a judicial verdict.

World Bank Estimates of Environmental Costs

Problem	*Impact*	*Low Estimate USD million*	*High Estimate USD million*
Urban Air Pollution	health	517	2102
Water Pollution	health (esp. diarrhoeal disease)	3076	8344
Soil Degradation	loss of agricultural output	1516	2368
Rangeland Degradation	loss of livestock carrying capacity	238	417
Deforestation	loss of sustainable timber supply	183	244
Tourism	decline in tourism revenues	142	283

Source: World Bank, 1995.

MONTREAL PROTOCOL

Depletion of Ozone layer is caused by emission of Ozone Depleting Substances (ODSs) particularly CFCs and halons. These result in increased ultra violet radiation which can seriously damage human health and environment. Global cooperation for the protection of the Ozone layer was institutionalised by the Vienna Convention in 1985 and the Montreal Protocol in 1987. The Montreal Protocol was designed to regulate the production and consumption of ODSs. The most common ODSs are CFCs which are primarily used in the refrigeration industry, aerosols propellants production and as surface treatment agents. The five most commonly used CFCs were to be phased out by 1996 in the developed countries. The protocol permits developing countries with an annual level of consumption of the controlled substances lower than 0.3 kg. to delay the phaseout by 10 years. The protocol also provides for financial support being made available for conversion to new substances and technologies. India falls under this category of low consumers and is, therefore, entitled to a longer phaseout period. The world bank has recently approved two important projects in India for the development of substitutes of ODS and related issues. However, the positive incentives of the Montreal Protocol in the form of technology transfer and aid have not come to India in any substantial form.

ISSUES IN WTO

The relationship between International Trade and Environment came up for discussion in Uruguay Round. Preceding that discussion, the International Community has already considered the need for adopting policies for sustainable development on a global basis. The Rio declaration on environment and Agenda 21 reflect these concerns. In fact, the Preamble of the Agreement establishing the World Trade Organisation (WTO) states that 'relations in the field of trade and economic endeavour should be conducted with a view to raising standards of living, ensuring full employment and a large and steadily growing volume of real income and effective demand, and expanding the production of and trade in goods and services, while allowing for the optimal use of the world's resources in accordance with the objective of sustainable development, seeking both to protect and preserve the environment and to enhance the means for doing so in a manner consistent with their respective needs and concerns at different levels of economic development'.

At the same time, it is emphasized that there should not be any policy contradiction between upholding and safeguarding a non-discriminatory and equitable multilateral trading system on the one hand and protection of the environment and promotion of sustainable development on the other.

At the time of approving the results of the Uruguay Round negotiations in Marrakesh in April 1994, the meeting also decided to initiate a comprehensive work programme on trade and environment in the WTO. The work programme is being organised by the WTO Committee on Trade and Environment (CTE). It has a broad work programme covering all areas of the multilateral trading system including goods, services and intellectual property. The CTE has both analytical and prescriptive mandate. It has to study the linkages between trade and environment measures with a view to promoting sustainable development and to make recommendations as to the need of any amendments in the current provisions of the multilateral trading system as enunciated in the WTO.

The report of the CTE was submitted to the WTO Ministerial Conference in Singapore in December 1996.

THE REPORT

The report recognises that there is a close link between poverty and environmental degradation. It is acknowledged that an open, equitable and non-discriminatory trading system, resulting in continued and expanding market access can help the developing countries obtain the resources to implement appropriate environmental policies, directed towards sustainable management of their environmental resources. It is also recognised that trade liberalisation by itself will not solve the problems of environmental degradation. Conversely, restricting trade will not help solve the problems of environment either. There are some studies, mostly carried out in the developed countries, which show that trade liberalisation under some conditions can help improve the environment. At a theoretical level higher environment standards in the developed countries can lead to resettlement of dirty industries in the developing countries where the environment standards are lower. Most studies, however, reveal that there is not much evidence to support this hypothesis. Especially, in the long run, if liberalised trade lead to higher income level in the developing countries, demand for environment-friendly products and processes will increase subsequently also in the developing countries. Therefore, it is possible that trade liberalisation can help improve aggregate environment standards globally in the long run.

INDIAN CONTEXT

However, it must be recognised that all types of industrialisation imposes certain environmental costs. For example, Silicon valley in USA contains the most polluted soil in America. The chemicals used to clean computer components are responsible for this environmental problem. Industrial activity imposes environmental costs on three accounts: first, it may result in degradation or exploitation of natural resources beyond a sustainable level such as, forest resources. Second, it may cause environmental pollution because of the production processes involved, such as leather tanning or dye-stuff production. Third, it entails costs in terms of health management, such as people in metropolitan areas, developing breathing problems due to air pollution. A recent World Bank study on India has estimated that the costs of environmental damage can be conservatively estimated at Rs. 340 billion annually. The table in earlier page provides some relevant data from this Report.

Though all industries are polluting to some extent, some are more polluting. The following industries have been identified as causing above average environmental pollution:

(a) Primary metallurgical producing industries including zinc, lead, copper, aluminium and steel
(b) Paper, pulp and newsprints
(c) Pesticides/Insecticides
(d) Refineries
(e) Ferilizers
(f) Paints
(g) Dyes
(h) Leather Tanning
(i) Rayon
(j) Sodium/Potassium Cyanide
(k) Basic Drugs
(I) Foundry
(m) Storage Batteries
(n) Acids/alkalies
(o) Plasters
(p) Rubber-Synthetics
(q) Cement
(r) Asbestos
(s) Fermentation industry
(t) Electroplating industry

Some of these industries are important for India's export trade, for example, leather dyes and intermediates and pharmaceuticals. In fact, cotton textiles which as a group account for about 30 per cent of India's exports is an industry which can cause substantial environmental damage. Cotton growing is highly pesticides-intensive. At the manufacturing stage, the textiles industry uses a large number of chemical dyes which have adverse environmental effects. Threat to India's exports due to environmental measures introduced by some of the developed countries is acute in the case of leather and textiles industry. It can also affect substantially agro-exports.

Some developed countries, with Germany in the lead, has banned the use of PCP in leather processing. PCP substitutes are available though these are costlier. While the switch to PCP substitutes reduces the cost competitiveness, the environmental impact is positive. There is a growing demand for eco-friendly textiles in the developed countries. Production of eco-friendly textiles requires minimum use of chemicals. A list of prohibited chemicals has been developed by the concerned agencies.

In a notification issued in May 1997, the Government banned the use of azo dyes in textiles. All these measures also have similar type of impact as in the case of leather sector.

In addition to the measures which are being introduced in response to the pressures being exerted by the developed countries, some action has been taken on the country's own initiative. For example, aqua-culture industry emerged as an important source of foreign exchange in the recent past. However, it can cause environmental damage in the coastal areas of Andhra Pradesh and Tamil Nadu where aqua-culture industry is mostly concentrated. Through a judicial verdict, further growth of aqua-culture has been brought to a halt. Tamil Nadu Government has passed an Act which stipulates that all new fisheries projects will have to be cleared by a district council, comprising the District Collector, pollution control board representative as well as other concerned officials. Similarly, a large number of tanneries in Tamil Nadu has been closed down through a court order because of environmental damage being caused by them.

There is no doubt that Indian Industry has to go for higher level of environmental standards in future both because the country's environmental protection needs such standards as well as the demand of the important trade partners.

2

Industrial Environmentalism and its Dimensions

RAKESH KUMAR AND MEERA SINGH

Man has been rapidly and deliberately exploiting the nature and natural resources, with the aid of science, industry and technology. No doubt industrialisation in certain respect is a bonaza for life, but evils accompanying it are also no less in number. The convenience of transport, mass-media, power generation, production of medicine, modern surgery, the clamour for modern gadgets, the pleasing colours of textiles etc. are a few good results of rapid industrial growth. But the most outstanding and patent danger that emanates from the industrial activities is pollution.

CONCEPTUAL FRAMEWORK

Human life is closely associated with the eco-system, which is the consequence of hundreds of millions of years of slow evolution of plants, chemicals and atmosphere.

Our environment mainly consists of two major elements:

(i) the physical environment comprising land, water and air (lethosphere, hydrosphere and atmosphere), and

(ii) the biological environment comprising all living organisms.

ORIGIN OF POLLUTION

There exist different views regarding the origin of pollution crisis on the planet earth. Certain authors such as Lynn White (1967) and John M.C. Harg (1969) blamed Judeo-Christian ethic for pollution. According to them this ethic taught man to believe that the earth was made for man and to do whatever he wished, and thereby encouraged exploitation. However, this view has been contradicted by Wright (1970). Wright pointed out that the Judeo-Christian religion teaches stewardship and he proposed that it is not religious belief, but human greed and ignorance which have allowed our culture to develop on ecological crisis like pollution.

Some authors (such as Southwick), associate the human population explosion with the pollution problems. They point out that with more people there is more sewage, more solid wastes, more fuel being burned, more fertilizers and insecticides being used to produce more food for hungry Meiths. But there are certain Writers who have pointed out that in under developed countries, pollution is not the severe problems as it is in technologically developed countries.

Finally, modern ecologists such as Odum, Southwick, Smith, etc. sought many factors such as human population exploison, unplanned urbanization and deforestation, profit oriented capitalism and technological advancement, (may have for originated) pollution crisis on earth. In fact, in countries where there have been the greatest technological advances, the worst pollution, occurs. Thus in well-developed nations on a per capita basis, citizens consume more food, use more pesticides, fertilizers, fuel, minerals, cars and other manufactured products of all kinds. Most of these products are manufactured in one or other kinds of industries, all of which in their turn add some pollutants in our environment and cause pollution.

TYPES OF POLLUTION

Generally pollution is classified either according to the environment (air, water, soil) in which it occurs, or according to the pollutant (lead, mercury, carbon di-oxide, solid waste, noise, biocide, heat etc.) by which pollution is caused. Sometimes pollution is also classified into two broad categories; (a) Natural pollution which originates from natural processes; and (b) Artificial pollution which originates due to the activities of man.

Industrial Pollution

The industrialisation is mainly concerned with physical environmental pollution (water, air and noise). Most of the Indian rivers and fresh-water streams are seriously polluted by industrial wastes or effluents of different industries such as Petro chemical complexes; fertilizer factories; oil refineries; pulp, paper, textile, sugar and steel mills, tanneries, distilleries, coal washeries, synthetic material plants for drugs, fibres, rubber, plastics, etc. The industrial wastes of these industries and mills include metals (copper, zinc, lead, mercury, etc.), detergents, petroleum, acids, alkalies, phenols, pesticides, alcohals, cyanide, arsenic, chlorine and many other inorganic and organic toxicants. All of these chemicals of industrial wastes are toxic to animals and may cause death or sublethal pathalogy of the liver, kidney, reproductive systems, respiratory systems, or nervous systems in both invertebrate and vertebrate aquatic animals (Wilbur). Chlorine which is added to water to control growth of algac and bacteria in the cooling system of power station, may persist in streams to cause mortality of plankton and fish. Heavy fish mortality in river Sone near Dehri-on-Sone in Bihar is reported to cause by free Chlorine contest of the chemical wastes discharged by factories near Mirzapur in U.P.

Mercury like other heavy metals such as lead and cadmium has cropped up as a toxic agent of serious nature. Mercury, a by product of the production of vinyl-chloride, is uesd in many chemical industries and it is also a by-product of some incineraters, power plants, laboratories and even hospitals (Aeranson). In Japan, illness and even death occurred in the 1950's among fishermen who ingested fish, crabs, and sheel-fish contaminated with Methyl Mercury from Japanese Coastal industries. This Mercury poisoning disease was called Minamata disease. Initial symptoms of Manamata disease included/ numbness of the limbs, lips and tongue impairment motor control, deafness and blurring of vision, cellular degeneration occurred in the cerebellum, midbrain, and cerebral Cortex and this led to spasticity, rigidity and coma.

The toxic and pathological effects of some heavy metal water pollutants have been tabulated in table:

Some Indian Rivers and their Major Sources of Pollution

Name of the River	*Sources of Pollution*
(1) Kali at Meerut (U.P.)	Sugar Mills; Distilleries, Paint, Soap, Rayon, Silk, Yarn, Tin and Glycerine Industries.
(2) Jamuna near Delhi	D.D.T. Factory, Sewage, Indraprashtha Power Station, Delhi.
(3) Ganga at Kanpur	Jute, Chemical, Metal and Surgical industries; tanneries, textile mills and great bulk of domestic sewage of highly organic nature.
(4) Gomati near Lucknow (U.P.)	Paper and Pulp Mills, Sewage
(5) Dajora in Bareilly (U.P.)	Synthetic rubber factories.
(6) Damodar between Bokaro and Panchet	Fertilizers, Fly ash from Steel mills, suspended coal particles from washeries, and thermal power station.
(7) Hooghly near Calcutta	Power stations paper, pulp, jute, textiles, chemical mills, paints, varnishes, metal steel hydrogenated vegetable, oils, rayon, and soap, match, shellac, and polythene industries and sewage.
(8) Sone at Dalmia Nagar (Bihar)	Cement, Pulp and Paper Mills.
(9) Bhadra (Karnataka)	Pulp, Paper and Steel Industries.
(10) Cooum, Adyar and Buckinghum Canal (Madras)	Domestic Sewage, Automobile workshops.
(11) Cauvery (Tamilnadu)	Sewage, tanneries, distilleries, Paper and Rayon mills.
(12) Godawari	Paper mills
(13) Siwan (Bihar)	Paper, Sulphur, Cement, Sugar Mills
(14) Kulu (Between Bombay and Kalyan)	Chemical factories, rayon, mills and tanneries.
(15) Suwao (in Balrampur)	Sugar industries.

Source: Environmental Biology, P.S. Verma and V.K. Agrawal.

Pathological Effects of Heavy Metal Water Pollutants on Man

Metal	*Pathological Effects on Man*
1. Mercury	Abdominal pain, headache, diarrhoea, hemalysis, chest pain.
2. Lead	Anaemia, Vomiting, loss of appetite, convulsions, damage of brain, liver and kidney.
3. Arsenic	Disturbed peripheral circulation, Mental disturbance, liver cirrhosis, hyperkeratosis, lung cancer, ulcers in gastro-intestinal tráct, kidney damage.
4. Cadmium	Diarrhoea, growth retradition, bone deformation, kidney damage, testicular atrophy, anaemia, injury of central nervous system and liver, hypertension.
5. Copper	Hypertension, uremia, coma, sporadic fever.
6. Barium	Excessive salivation, vomiting, diarrhoea, paralysis, colic pain.
7. Zinc	Vomiting, renal damage, cramps.
8. Selenium	Damage of liver, kidney and spleen, fever, nervousness, vomiting, low blood pressure, blindness, and even death.
9. Hexavalent chromium	Nephritis, gastro-intestinal unceration, diseases in central nervous system, cancer.
10. Cobolt	Diarrhoea, low blood pressure, lung irritation, bona deformities, paralysis.

Source: Environmental Biology, P.S. Verma and V.K. Agrawal.

Water Pollution

Water is essential ingredient of life. Human body itself has by weight 70 percent water. One billion world's population is compelled to drink dirty water. According to National Environmental Engineering Research Institute Scientists 70 percent of India's inland water is unfit for human consumption. Factories discharge its ragged, sluttish harmful scrap of waste husk materials, ash effluents, ails and chemical into nearby rivers, lakes, sea, ponds and wells. Water pollution also occurs due to toxic as well as radio-active materials. All the rivers and seas are facing the problem of water pollution. Only in Hugli at Calcutta untreated water of different 160 industries is discharged in it causing serious water pollution. Due to pouring of effluents of different leather textile and other industrial units situated at Kanpur, the most industrialised city of U.P. Ganges water becomes black at Kanpur.

From Gangotri to Varanasi, at least 1611 rivulets fell down in Ganges with a huge amount of sewage, untreated effluents and industrial wastes.

According to an estimate, two-thirds of all illness in our nation are related to water borne diseases such as diarrhoea, typhoid, jaundice, cholera, dysentry etc. Many of which have assumed epidemic dimensions. The World Health Organisation's study reports that 80 per cent diseases are caused due to lack of proper sanitation and short of drinking water. A survey opinion is that 73 million work-days are lost annually to water related diseases, the cost of treating them and the loss in production is amounted to Rs. 600 Crore per annum in India. A lot of amount is being spent by the Government of India to remove water pollution e.g., in the seventh plan, Rs. 250 Crore was spent to cleanse the polluted Ganga. Industrial waste should be well treated on the industrial complexes and then poured in the run off water. Central Board for the Prevention and Control of Water Pollution is making efforts for reducing and checking the water pollution throughout the country. Thanks to our Government for establishing Central Ganga Authority (C.G.A.) meant for cleansing of Ganga water but proper utilisation of funds is needed and this scheme should be extended to other rivers.

Water pollution may be classified into surface water pollution, ground water pollution, soil water pollution, etc.; also as river pollution, lake pollution, esturine pollution, coastal water pollution, open ocean pollution, etc. Based on the source or type of contamination, it may be nutrient pollution, bacterial pollution, viral pollution, metallic pollution, petrochemical pollution, pesticide pollution, thermal pollution, radio active pollution, etc.

These aquatic pollutants come from many sources. Excessive nutrients, such as nitrates and phosphates, commonly originate in domestic sewage, run-off from agricultural-fertilizer. These nutrients cause pollution primarily because they stimulate the growth of micro-organisms which often increase the biological oxygen demand (BOD) of the water and reduce the dissolved oxygen available for fish, higher animals and other aquatic animals, Toxic chemicals as agents of water pollution originate from industrial operations. However, Kimball (1975) recognized only three major sources of pollutations; domestic, industrial and agricultural origin.

Ecology of Water Pollution

Each type of water pollution affects the abiotic and biotic factors of different aquatic systems in different degrees and ultimately effects man.

(1) Sewage Pollution

Contamination of fresh-waters and shallow offshore seas by sewage is a common occurrence. Domestic sewage and waste water is about 99.9 percent water and 0.02-0.04 percent solids of which proteins and carbohydrates each comprise 40-50 percent and fats 5-10 percent (Simmais). In other words, sewage includes mostly biodegradeable pollutants such as human faecal matter, animal wastes, and certain dissolved organic compounds (e.g. carbohydrates, urea, etc.) and inorganic salt such as nitrates and phosphates of detergents and sodium, potassium, calcium and chloride ions. Under natural processes most of the biodegradeable pollutants of sewage are rapidly decomposed, but when they accumulate in large quantitites, they create problems. Most cities of well developed countries like U.S.A., Britain, etc. and some cities of developing countries like India have evolved certain engineering systems, such as, septic tanks, oxidation ponds, filter beds, waste water treatment plants and municipal sewage treatment plants for the removal of many harmful bacteria and other microbes, organic wastes and other pollutants from the sewage, before it is tipped into river or sea.

Sewage treatment is usually performed in following three stages:

(I) Primary treatment, which removes large objects and suspended undissolved solids of raw sewage and converts them into a biologically inactive and aesthetically inoffensive state, the slugde, a valuable fertilizer;

(II) Secondary treatment, which supplies aeration and bacterial action to decompose organic compounds into harmless substances such as CO_2 Sulphate and water during later stage of secondary treatment, whole waste water is chlorinated (i.e., terated with chlorine) to reduce its content of bacteria;

(III) Tertiary treatment, which removes nitrates and phosphates and release pure water.

These three stage of sewage treatment have become increasingly expensive and only in most advanced countries all the three treatments of sewage or waste-water treatment plant or have inadequate sewage treatment facilities. Consequently, normally and especially during heavy downpour and floods, raw sewage or incompletely treated sewage is dumped into rivers which cause severe 'water pollution problems in following ways:

(i) Bacterial and viral contamination; and
(ii) Eutrophication.

(2) Thermal Pollution

Various industrial processes may utilise water for cooling, and resultant warm water has often been discharged into streams or lakes. Coal-oil-fired generators and atomic energy plants generate waste heat which is carried away as hot water and cause thermal pollution. Thermal pollution produces distinct changes in aquatic biota. A temperature rise of 10°C will double the rate of many chemical reactions in the body and exert a disruptive effect on aquatic ecosystems.

(3) Silt Pollution

A result of intensive agriculture, earth moving for construction projects, poor conservation practices and down pour with resultant floods, increases production of silt in stream and lakes. This load to particulate matter cuts down primary productivity by decreasing the depth of light pentration. Silt may also interrupt or prevent the reproduction of fish, by smothering eggs laid on the bottom.

(4) Estuarine and Oceanic Pollution

Often oceans are considered as so vast that they are virtually unlimited in their ability to accomodate the waste products of human civilization. But there are substantial evidence to indicate global pollution of coastal waters and open oceans due to dumping of domestic and industrial wastes, sewage, oil drilling in coastal waters, spilling of oil form tankers etc. The oceans have in fact, become the final settling basin for millions of tons of waste products from human activities. For example in the late 1960's West Germany was dumping 375 tons of Sulphuric acid, 750 tons of iron sulphate, 10 tons of chlorinated hydrocarbons and 16,000 tons of gypsum wastes into the North sea and North Atlantic every day. Table shows a few of the industrial and agricultural pollutants reaching the ocean annually in the mid 1970s.

Examples of Industrial and Agricultural Pollutants discharged Annualy in to the World's Oceans (after Southwick, 1976)

	Pollutants	*Estimated Annual discharge, 1970-75 in Metric tonnes*	*Source*
1.	Petroleum and industrial hydro-carbons	3,405,000	Off shore wells, oil tankers, industrial wastes.
2.	Hydrocarbons (Airborne)	15,000,000	Vehicles, Industries, power plants.
3.	Airborne lead	350,000	Vehicles.
4.	Mercury	100,000	Industrial Operation
5.	Aldrin-Tonaphene (Converted to dieldrin)	25,000	Agricultural and public health operations.
6.	Benzene hexachloride	50,000	-do-
7.	D.D.T.	25,000	-do-
8.	Palychlorinated biphoneils (Pecs)	25,000	Plastic industries

Source: Environmental Biology, P.S. Verma and V.K. Agrawal.

This tremendous burden of pollutants is evidently affecting the health and integrity of the world's oceans. Due to oceanic pollution the Marine biota has been seriously affected. Courteau reported a 30 to 40 percent decline in the over all productivity of pelagic organisms from Shrimps to Whales in the past 25 years. He also observed a serious Shrinkage of coral reefs in many trophical areas of world.

There has occurred a decline in populations of many fishes due to oceanic pollution. Oil spills have killed water birds, mammals, fish and vegetation (Hunt).

Control of Water Pollution

The intensity of water pollution can be minimised by following methods:

1. Adequate Sewage Treatment

Raw sewage should not be dumped in rivers or oceans. Before its disposal into them, sewage should be properly treated in sewage treatment plants.

2. Treatment of Industrial Effluents

The industrial effluents should be cleaned before they are discharged into rivers.

3. Recycling

The best method of prevention and control of water pollution is the recycling of various kinds of wastes. Dung of cow and buffaloes can be used for the production of gobar gas, a cheap source of fuel and also of manure.

Air Pollution

The air is also polluted by smoke and fog (smog) dust and fume, which are emitted by the industrial concerns—Today clean air is a rare commodity in overcrowded big cities and towns. If water is defiled, one may drink it after purification, but we can not avoid breathing polluted air, we have to take it as it comes. One more problem associated with air pollution is that it is invisible which can be hardly seen by microscopic test only.

Industries produce as by-products poisonous gases and oxides which are puffed out into the atmosphere. Various dangerous gases as carbon mono-oxide, nitrogen exides, chlorine, ammonia, sulphur dioxide, hydrogen sulphide, methane; can be detected only by special instruments. The unfortunate Bhopal Catastrophe in which flow of Methyl-isocynate (MIC) from Union Carbide Corporation Ltd. (UCCL) took the life of at least 2,500 people, is the most repulsive and tragic accident in the industrial history of the world. Scientists have developed a vide range of equipments to curb air pollution. Technological capabilities and control devices are available to abate air pollution problems, though same may not yet become economically-feasible, specific legislative controls are being regulated to restrict air pollution caused by industrial misconduct.

A major source of air pollution is the particulate and gasous mateer released by the burning of fossil fuels such as coal, petroleum, etc. Out of this comes a variety of emissions: (1) Fine particles—

which include carbon particles, metallic dusts, tors, resins, aerosols, solid oxides, nitrates and sulphates; (2) Coarser particles—largely carbon particles and heavy dust that is quickly removed by gravity from the air; (3) Sulphur compounds; (4) Nitrogen compounds; (5) Oxygen compounds; (6) Halogens; and (7) Radio-active substances.

These pollutants are artificial pollutants and they are poured in air mainly by at least five major fuel burning sources. The transportation industry including Automobiles (cars, scooters, torcycles) is the greatest sources of air pollution. They produce nearly two-thirds of the carbon mono-oxide and one-half of the hydrocarbons and nitrous oxides. The automobile exhaust also contains leaded gas and particulate lead. Electrical power plants burning fossil fuels, particularly coal and sometimes petrol or diesel, produce two-thirds of the sulphur dioxides, Industrial processers such as, Metallurgical plants and smelters, chemical plants, petroleum refineries pulp and paper mills, sugar mills, cotton mills and synthetic rubber manufacturing plants are responsible for about one fifth of the air pollution. Heating plants for homes, apartments, schools and industrial buildings are the fourth largest source of air pollution.

Other sources of air pollution are minor in quantitites but bear significance due to the harmful substances they release. They are agriculture, which is responsible for pesticides, dust from agricultural practices and field burning and the construction industry. Nature too adds few natural pollutants such as pollen, hydrocarbons released by vegetation, dusts from deserts, storms, and volcanic activity.

Some of the most common air pollutants, their sources and their effects on human health have been tabulated in table.

To measure and control the magnitude of air pollution in various industrial centres of India, National Environmental Engineering Research Institute (NEERI) has set air monitoring stations in Bombay, Calcutta, Delhi, Madras, Hyderabad, Kanpur, Jaipur, Ahmedabad and Nagpur. In one of the survey conducted by NEERI in 1970 to measure the air pollution by Sulphur dioxide and suspended particles in some major cities of India, it is found that Chembur-Trombay area of Bombay has highest pollution, while, New Delhi has highest air pollution of suspended particulate matter. In another survey, Calcutta is reported to have highest carbon mono-oxide pollution during peak traffic hours. dust collectors. The particulate pollution is measured by the instrument

METHOD OF DETECTION AND MEASUREMENT OF AIR POLLUTION

Air pollution is usually measured by sampling of air by thermal and by electrostatic precipitation sonkin impactor and electrostatic.

Common Air Pollutants, their Sources and Pathological Effects on Men (After Southwick, 1976)

	Pollutants	*Where they come from (Source)*	*Pathological effect on men*
1.	Aldehydes	Thermal decomposition of fats, oil or glyceral	Irritate nasal and respiratory tracts
2.	Ammonias	Chemical processes-dye making; explosives, fertilizer	Inflame upper respiratory passages
3.	Arsines	Proesses involving metals or acids containing arsenic soldering	Break down red cells in blood damage kidneys cause jaundice.
4.	Carbon Mono oxides	Gasalone, motor exhausts, burning of coal	Reduce oxygen carrying capacity of blood.
5.	Chlorines	Bleaching cotton and flour; many other chemical processes	Attack entire respiratory tract and Mucous, Membrances of eyes; cause pulmonary edema.
6.	Hydrogen cyanides	Fumigation; blast furnaces, chemical manufacturing metal plating	Interfere with nerve cells, produce dry throat indistinct vision, headache
7.	Hydrogen Flourides	Petroleum refining glass etching; aluminium and fertilizer production	Irritate and corrode all body passages
8.	Hydrogen sulphides	Refineries and chemical industries; bituminous fuels bituminous fuels	Smell like rotten eggs; cause nausea, irritate eyes and throat
9.	Nitrogen oxides	Motor vehicle exhausts soft coal	Inhibit ciliary action so that smoke and dust penetrate into the lungs.
10.	Phosphoenes (Carbonyl chloride)	Chemical and dye manufacturing	Induce coughing, irritation and sometimes pulmonary edema.
11.	Sulphur dioxides	Coal and oil combustion	Cause chest constriction headache vomiting, and death from respiratory elements.
12.	Suspended particles (ash, soot, smoke)	Incinerators, almost any manufacturing	cause, eye irritations and possibly cancer

Source: Environmental Biology, P.S. Verma and V.K. Agrawal.

called deposit gauge or by Owen's dust counter. The thickness of the smoke is measured by Liegeon Sphere and by Ringelion Chart. The rough estimation of SO_2 in air can be made by chemical analysis of the dust collected in a deposit gauge or by a bubbler method fluorides are estimated by colour reactions.

Levels of Sulphur Di Oxide and Suspended Particulate Matter in Air During 1970 in Some Indian Cities

(From Sept., 1976)

	City	*Mean value of SO_4 Microgram/cubic metre*	*Suspended particulate matter Microgram/cubic metre*
1.	Bombay	47.1	240.8
2.	New Delhi	41.4	601.1
3.	Calcutta	32.9	340.7
4.	Kanpur	15.9	543.5
5.	Ahmedabad	10.7	306.6
6.	Madras	8.3	100.9
7.	Nagpur	7.7	261.6
8.	Hyderabad	5.1	146.2
9.	Jaipur	4.2	146.1

Source: Environmental Biology, P.S. Verma and V.K. Agrawal.

Ecology of Air Pollution

Once injected into the atmosphere, pollutants enter the biogeochemical cycles by different routes.

But if air masses over cities become stagnant, pollutants accumulate quickly and deteriorate air quality which causes many respiratory diseases in man and other animals. Air pollutants also accumulate during temperature conversions, when cooler surface layers of air become trapped under warmer upper layers. In these situations, the upper layers of warm air prevent the vertical rise and dispersal of pollutants which are held near the ground. Temperature inversions, commonly occur in cities surrounded by mountains or bordered by mountains on the leeward side.

Further a portion of air pollutants reaches land as dry fallouts;

it may then enter various nutrient cycles and food chains through water and soil. Other contaminants of air react chemically or photo-chemically with each other and produce such secondary pollutants as sulphuric acid, ozone, and peronyaectyl nitrate or PAN. Aerosols, and other forms of fine particulate matter act as condensation nucleus, to which water vapours present in the air quickly surround to form droplets of fog or rain.

Most kinds of air pollutions can be controlled by modern technology, but the costs ultimately be borne by the public in the form of higher prices for manufactured goods, higher taxes, reduced profit margins in industry, and more restrictions on individual activities such as burning leaves and trash and use of automobiles. The benefits involve not only improved environmental quality, but improved health, improve environmental quality, improved agriculture and plant growth and reduced deterioration of material goods.

In U.S.A., various devices such as the positive crenkease ventilation valve and catalytic converter have been developed to reduce exhaust emissions by automobiles, but these devices are not always fully maintained by the public. Likewise particulate pollution from industry and power generation can be controlled by electrostatic precipitators which are capable of dramatically reducing smoke and dust. Gaseous pollutants of industry and power station can be removed by chemical means, i.e., differential salubility of gases in water. A fine spray of water in a device known as a "Scrubber" can effectively separate many gases such as Ammonia and Sulphur Dioxide. Other gases may be removed by filtration or absorption through activated carbon, and still others by chemical conversion to exert or innocuous materials.

Legal Control of Air Pollution

In India air pollution control legislation envisages the formation of air pollution boards at the Central and State levels "With powers to issue and revoke licences to polluting Industries, enforce emission standards and frame rules and regulations for the control of air pollution". The legislation is primarily directed at the highly polluting industries such as iron and steel, textiles and power plants. The Board will have power to prohibit certain trades and manufacturing processes in notified aresa and prescribes emission standards in scheduled premises. The legislation is also understood to ban the burning of garbage and other waste products in urban areas as well as the faulting up of air by burning smoking fuels for domestic purposes. Recently, Bombay Smoke Nuisance Act has been enforced only in case of smoke emanating from the Chimneys of Industrial units. However, there is a need of legislation to deal with fumes of petro chemical units, ash, carbon particles, unpalatable smell and even noise from industrial units. (See Gopal Bhargava, in the *Hindustan Times Magazine*).

Noise Pollution

Now "Noise pollution" is an addition in the world of pollutions. In the recent year industrial noise is regarded a serious problem for mankind. Early morning we get up and hear a lot of noise. In the night we are not able to get a sound sleep because of various noise caused by different factories. In our daily life noise arises from big plants, compels us to shut the ears, and it has adverse effects on hearings and blood pressure resulted in heart diseases. Due to noise pollution many children are born deaf now a days. For measuring the sound decibel (D.B.) unit is used. According to the studies made by the scientists, it is said that a man may become deaf living in a atmosphere with a noise above 85 D.B. Noise above 120 D.B. imposes reverse effects to the pregnant ladies and their newly born babies. The industrialisation oriented hullabaloo has made difficult to good sleeping and dwelling. The industrial development since the advent of Industrial Revolution, is sundering the precious human peace and life.

High intensity sound or noise pollution is caused by many machines man has invented during his technological advancement. Thus, there exists a long list of sources of noise pollution including different machines of numerous factories, industries and mills, different kind of auto and motor vehicles such as scooters, motor bikes, cars, tempos, buses, trucks, tractors, aircrafts motorboats, ships, loudspeakers, special gatherings, loud pop-music, supersonic air-crafts and others.

Effects of Noise Pollution

Noise pollution has certain well evident ecological and pathological effects on bio and human beings.

There is a clear evidence now that the hair cells of organ of corti of inner ear can be permanently damaged if they are subjected to repeated sounds of high intensity before they have an opportunity to recover. Workers of different industries develop noise-deafness. Rok Music (and Disco Music) too is sound to produce temporary deafness in the listeners and also in the members of rock band. Further, noise is found to increase the level of Cholesteral in the blood, to increase blood pressure, and to cause headache. Noise pollution also causes the pupils of the eyes to dilate and mental abnormalities in foetus, hypertension, peptic ulcers andultimately, may lead to emotional and behavioural problems in man.

Recently many nations have enacted laws to penalise noise production by the vehicles or any type of industry.

INDIAN CONSTITUTION

Government of India (State) has made constitutional provisions for environmental management. The Government of India will look after the environment not only on the instance of the Courts when an individual is seeing the enforcement of his right, but also under specific provisions of Indian Constitution:

(a) Article 42. The State should make provisions for just and human conditions of work,

(b) Article 43. Securing living wage is not enough. State should endeavour to ensure decent standard of life,

(c) Article 47. State to raise the level of nutrition and standard of living and to improve proper health,

(d) Article 48-A. State shall endeavour to protect and improve the environment and to safeguard the forests and wild-life of the country.

(e) Article 51-A (g). It is fundamental duty of every citizen of India to protect and improve the national environment.

Industrial growth during Post-Second World war period is coming forth to environmental disbalance and industrial hazard. The problems of industrial environmental degradation is multipronged of which can be no single solution. For maintaining ecological balance some new sophisticated safety measures should be included in our Statute Book as the Factories Act, 1948, the Insecticides Act, 1968, the Water (Prevention and Control of Pollution) Act, 1974, The Air (Prevention and Control of Pollution) Act, 1981 and the Environments Protection

Act, 1986. But mere legislation can never be enough if it is not enforced strictly. Our environmentalists should vigilantly see the enforcement of the anti-pollution laws so that profit-minded industrialists should be compelled to adopt modern scientific techniques to minimise the pollution and noise pollution 'Back to nature' should be the slogan in the mind of people for protecting and survival of mankind and its health.

Notes and References

1. M.R. Garg and N.S. Tiwana, Enforcement and Environmental Law and Management of Pollution Control, in the book edited by R.K. Sapru, Environment Management in India, Ashish Publishing House, New Delhi, 1987.
2. H.K. Singh, Environmentalism and Social Responsibilities of Businessmen, *The Commerce Journal*, Allahabad, Vol. XXXV, 1990-91.
3. N.W. Chemberlain, Enterprise and Environment—The firm in time and place, McGrow Hill Book Co., New York, 1981.
4. P.S. Verma and V.K. Agrawal, Environmental Biology, S. Chand & Company Ltd., 1996.
5. A.S. Azad and Others, Nature and Extent of Heavy Metal Pollution Emanating from Industrial Units in Ludhiana, Seminar Proceedings of Punjab Agricultural University, Ludhiana, 4 Feb. 1985.
6. Biman Basu, Environmental Protection; A Many-Faceted Problem, Current Topics, Ambala Cantt. July 1986.
7. H.K. Singh and Meera Singh, Frontiers of General Management, Kanishka Publishers, Delhi, 1995.

3

Mineral Development and Environment

S.C. LAHIRY

India produces as many as 64 minerals: 4 fuel minerals, 11 metallic minerals and 49 non-metallic industrial minerals. There were 3794 operating mines in 1993-94 (provisionally) in the country excluding petroleum and natural gas wells, mines of 'minor' minerals and atomic minerals, coal and lignite accounted for 562 mines, 664 mines were metalic mineral mines and the rest were non-metallic mineral. The total value of minerals production in the country was Rs. 26386.5 crore in 1993-94. The distribution of value of mineral production in 1993-94 showed that fuels accounted for 84.7%, metallic minerals 7%, non-metallic minerals 3.6% and minor minerals 4.6%. Almost the entire value of metallic mineral production was shared by iron ore, chromite, copper ore, manganese ore, bauxite, etc. Among non-metallic minerals 89% of the value was shared by limestone, apetite, dolomite, kaoline, barytes, etc.

Mining industry being site specific and located in tribal and interior areas, provides employment to local people directly in mines. In fact mining activity can be a vehicle in uplift of socio-economic

This article first appeared in Yojana, Volume 41: 8, August 1997. Reprinted with permission.

status of the local people. Indian mining industry is characterised by a large number of small mines. In line with the changes in economic policies as brought out in July, 1991 by the Government of India, the new National Mineral Policy was announced in 1993. This has opened substantially the gates for private investment both domestic as well as foreign in Indian Mineral Industry. This is expected to give a fillip to the domestic mineral industry in furthering its growth and making an entire range of minerals and mineral based products internationally competitive.

The scale and level of requirement of minerals have increased manifold in our country and it is heading towards the stage where much larger consumption of mineral will be inevitable to sustain even the minimum growth rate of our economy. It is pertinent to note that out of the total land area of the country (3.29 million sq. kms.), the area leased out of mining, as on 1.1.94, was 7126.13 sq. kms. comprising about 9,213 mining leases excluding atomic minerals, minor minerals, petroleum and natural gas. This constitutes only about 0.25% of the geographic area of the country and that including atomic minerals and minor minerals it may be around 0.28% of the total area. Although area occupied for mining activity is small yet the damage to the environment on account of mining is causing grave concern. Environmental degradation resulting from mining activity in general can be briefly enumerated as follows:

(i) Air pollution with dust and gases due to drilling, blasting, mine haulage and transportation by road, and also from waste heaps;
(ii) Water pollution when atomic elements and other harmful elements are present in the ore/mineral mines effluents;
(iii) Modifying water regime such as surface flow, ground water availability and lowering down of water table;
(iv) Soil erosion, soil modification with dust and salt;
(v) Noise and vibration problem in mine and adjoining habitat including wild life;
(vi) Alteration of the land form;
(vii) Deforestation affecting flora and fauna; and
(viii) Spoiling aesthetics with untreated waste dumps.

Legal Framework

In order to harmonise mineral development with environment, the mining sector is regulated by the Environment (Protection) Act, 1986 (EPA), the Forest Conservation Act, 1980, the MMRD Act, Wild Life Act, 1972 (Prevention and Control of Pollution) Act, 1974 and Air (Prevention and Control of Pollution) Act, 1981.

For exploration and mining on forest land, prior permission of the Government is required under the provisions of the Forest Conservation Act, 1980. EPA and the Rules framed thereunder, provide for the prospecting and mining activities only with prior permission from the Ministry of Environment and Forests for all mining projects over 5 hectares of area or in all cases in which an existing mining lease in excess of 5 hectare in area is proposed to be further expanded. Site clearance and environment clearance are also required in accordance with guidelines issued under the EPA. For this purpose, detailed baseline information, Environment Impact Assessment (EIA) and Environment Management Plan (EMP) are required to be submitted by the mine owners.

The Government in exercise of its power under MMRD Act, has made comprehensive provisions under the Mineral Conservation and Development Rules, 1988, for environmental protection. The licensee/lessee is duty-bound under the Mineral Conservation and Development Rules to prepare and get approved a mining plan which inter alia prescribes action to take precautions regarding:

(a) Removal and storage of the top soil, over burden waste and sub-grade material.
(b) Reclamation and rehabilitation of lands;
(c) Precaution against ground vibrations and fly rocks;
(d) Precaution against air pollution;
(e) Precaution against polluting material water courses around the mining areas;
(f) Discharge of toxic fluids/tailings;
(g) Precaution against noise, and
(h) Restoration of flora and fauna.

The National Mineral Policy 1993 for non-fuel and non-atomic minerals, prohibits mining operations in identified ecologically fragile and biologically rich areas and strip mining in forest areas as far as possible. The latter could be permitted only when accompanied by a comprehensive time bound reclamation programme. It states further that no mining lease would be granted to any party, private or public, without a prior mining plan including the environmental management plan approved and enforced by statutory authorities. The environmental management plan should have adequate measures for minimising the environmental damage, restoration of mined areas and for planting of trees in accordance with the prescribed norms. As far as possible, reclamation and afforestation will proceed concurrently with mineral extraction. Efforts should also be made to convert old mining sites into forests and other forms of land use.

PROBLEMS

The impact of mining in environmental degradation has been receiving attention in recent years. An overwhelmingly large number of small mines being open pit mines, land degradation including deforestation takes place consequent to mineral exploitation. Where the extent of damge caused by mining may not be significant individually the cumulative effect, specially by a cluster of mines, becomes fairly large. Air pollution is quite common due to blasting and transportation of minerals and very largely by crushing units where such crushing is essential for marketing the mineral. Even unstabilised waste dumps cause air pollution and wash off during monsoon. Overburden and waste generated during mining, degrade land and can pollute adjoining farmlands and water sources.

The notification issued in January 1994 (as amended in May 1994) under the Environment (Protection) Act and Rules made thereunder requires all mines (major minerals) having a mining lease area of more than 5 hectares or in every case in which an existing mining lease in excess of 5 hectare in area is proposed to be further expanded, need to get a clearance from the Ministry of Environment and Forests. For this purpose a detailed application supported with mining feasibility report. EIA baseline and analysis data, EMP and a filled in questionnaire are required to be submitted.

On the issue of plantation and afforestation in mines, in small lease hold areas unless a good part of the area is left unmined, it is not possible to plant large numbers of saplings every year. In small lease holds greenery development should not be taken up in peripheral areas after initial benches have been mined out. At a later date the pits can be filled up for reclamation and afforested.

As already mentioned India is blessed with a significant number of large and small manual and mechanised mines. From the modest beginning, development of large open cast mining projects for iron ore, limestone, bauxite, etc. took place. In the succeeding paragraph, the impact of development efforts, environment management aspect, result achieved, etc. for the aforementioned minerals alongwith export potential mineral marble in India are discussed:

Iron Ore

There are about 257 iron ore producing mines in the country. Madhya Pradesh accounts for about 26%, followed by Goa with 22%, Bihar 19% and Karnataka 18.8% are amongst main iron-ore producing states. Production of iron ore (including concentrates) during the year 1996-97 is estimated to be 69.1 MT. Out of estimated despatches of iron ore (63.8 MT), the share of despatches of iron ore for internal consumption would be 40.4 MT and the balance is for export.

The internal consumption continue to record significant growth because a number of new capacities are coming up in the country. According to a study of the world iron ore industry by the Brussels-based International Iron Steel Institute (IISI), a production of over 86 MT of iron ore is envisaged in 2005. From a level of 27 MT, iron ore export in 1995 is expected to rise to 37 MT in the year 2000. The Indian iron ore industry has the export capability of over 30 MT per year at current level of off take. For the year 2000, the major Indian exports are expected to be Kudremukh with 10 MT pa, NMDC (Bailadila + Donnimalai) with 7.4 MT, pa, Goa shippers 18 MT, while producers in Bihar, Orissa and Bellary Hospet, etc. account for an additional 2.5 MT. Last year Goa accounted for 52% of the country's iron ore exports. The GSI survey indicates that given the current intensity of mining, Goa's ore reserves are expected to peter out by 2020 A.D.

The State came into focus recently, when Goa's valuable pre-historic sites were being badly affected by the pollution. Though mining is the biggest contributor to the State's economy, the degradation of the ecology was a matter of great concern because the best of efforts had failed to combat ore rejects from flowing into the rivers and affecting marine life and potable water. For every tonne of ore extracted, two and a half tonnes of waste are created. Many mining areas have been abandoned after extracting the ore, leaving behind large craters.

Mining companies do take precautionary measures to control the flow of mining rejects, but such measures are not enough.

Agricultural land located close to the reject dumps has been adversely affected. In this context, some of the environmental control measures adopted by National Mineral Development Corporation (NMDC) in their Bailadila Iron Ore mines are worth mentioning. It operates the largest mechanised iron ore mines in the country at Bailadila (MP) and Donnimalai (Karnataka). Apart from maintaining ambient air quality standards in the work zone of the iron ore projects, flow of the suspended solids from the screening plants of respective mines, or run-offs is being controlled by having tailing dams constituted of proper designs. Besides a number of check dams have been constructed to arrest fines and reduce the velocity of flow even before reaching tailing dams. At the upstream of the tailing dam, conditions are created for discrete and quiescent setting of solids. The quality of water discharged from the tailing dams, which act as the pollution control structure is observed to meet the standards prescribed in GSR 422E. In the absence of proper pollution control measures, the run-offs of Bailadila mines created heavy pollution which made streams and river Shankhani unfit for human consumption. Tailing dam function as a pollution control facility where iron ore slime settles by physical means and clear water discharges downstream. Since the dam is new, recycling of solid waste is yet to start. The waste dumps currently being piled up would be reclaimed after they reach their peak accommodating capacities. Systematic afforestation is underway. About 16 lakh saplings have been planted so far and the survival rate is found to be encouraging (+90%). Deposit number 5 of Bailadila is surrounded by lush greenery. Donnimalai iron ore project has been adjudged the best opencast mechanised mine in India (instituted by FIMI).

MAJOR CAUSE

In the iron ore and limestone mines, a major cause of air pollution is transporting huge volume of iron ore, limestone, etc. through open trucks/dumpers etc. It also leads to very heavy consumption of diesel, fuel oil, etc. Burning of diesel/fuel oil, etc. and transportation of minerals combining together aggravate dust, smoke and noise pollution in the mines. Attempts are being made the world over to develop possible alternatives systems for surface transportation in mines in varying operating conditions. Transportation systems e.g. belt conveyer and slurry transport are therefore being increasingly under in USA, Canada, etc. The systems are reliable and provide higher availability. In Goa iron ore area, a belt conveyer system operating successfully in a 3.5 MT pa mine for transportation of ROM from quarry bottom involving a lift of around 300 metres. Slurry transport has been used in a large iron ore mine. The use of these technologies in mines has become an alternative to trucks for ore and over burden transportation.

Limestone

India is endowed with abundant resources of limestone. The total reserve is of the order of 76,440 MT. Production of limestone is expected to reach 98 MT during 1996-97. Cement industry consumes 88.5% of limestone followed by iron & steel sector 6%, chemicals industry 3%; and so on. Limestone mining is carried out entirely by open-pit method in small, medium and large scale. Limestone and iron ore account for about 70% of the total number of opencast mechanised non-coal mines. Technologies adopted in open-pit mines in India are generally comparable to those prevalent in other countries for the size of the deposits and the scale of operations.

Following are the environmental impact areas relating to typical limestone mining (a) land degradation, (b) removal of top soil, (c) disposal of solid waste, (d) disposal of water regime and drainage pattern, (e) air quality, (f) noise pollution, (g) ground vibrations, (h) socio-economic change, etc.

The abatement measures being carried out in limestone mining for controlling damage to the environment are as follows:

The reclamation of mined land is done by land-scaping or site preparation, soil amelioration and re-vegetation. Even though there is large gap between the area degraded by mining and that reclaimed so far, considerable emphasis is being laid on the need for back filling of the worked out area with waste so that they could be reclaimed. With respect to air pollution, suppression of dust by water sprays is one of the measures very commonly used in Indian mines—be it drilling, blasting, or that generated in road haul. Dust extractors are used in large diameter drilling in mechanised open cast mines. Control of water pollution is done by provisioning of garland of drains and check dams etc. which are some of the common abatement measures towards checking of water pollution. Control of noise pollution is achieved by (i) reducing sound at source, (ii) interrupting the path of noise, (iii) reducing ground vibrations, etc. It has been observed that some limestone quarry after exploitation remain unattended and develop crater and subsidence in the absence of rehabilitation measures. The disfigurement of the mining area is very common; almost barren mined area is also a common sight.

Bauxite

Seventy parcent of the world bauxite resources are situated in the developing countries including India. India ranks sixth in the world bauxite deposits (7.5% share). World bauxite production is of the order of 125 MT whereas India's production is around 5 MT. India has adequate resources which need to be developed early. Two bauxite mines with export oriented aluminium plants are coming up. Still there is scope for further bauxite mining for export oriented alumina plants. The resource position indicate a quite comfortable position for India for at least 50 years. The eastern ghat region accounts for more than 90% of country's metallurgical grade bauxite reserves. In 1995-96 and 1996-97, bauxite production was around 5.1 MT. There are 224 mines, most of them small, opencast and manually operated, except 15 major mines which account 72% of production and of this one mine, i.e. Panchapatmali mine accounts for 45% of country's total production. Orissa is the leading producer followed by Bihar, Gujarat, M.P. etc. in bauxite.

In India, bauxite exploitation is done by open-cost mining method only. Mining machinery and processed industrial materials e.g. explosives for blasting are mainly required to win lateritic bauxite by open cast method. Like any surface mining, bauxite exploitation causes definite disfigurement of earth surfaces and reddish brown patches appear on the plateau top. Major environmental problems associated with bauxite mining are deforestation, removal of top soil, alteration of landscape, surface and ground water contamination, air, dust and noise pollution and generation of waste products. However, an effective environmentai management plan can minimise the ecological disturbances and even the bauxite mining site may be transformed into lush green forest after few years of mining. Thus a bauxite mining venture in the ecologically sensitive areas becomes more cumbersome and costly. Some large and viable bauxite deposits of Eastern Ghats are located in ecologically sensitive areas with thick forest cover and mining in some cases would become prohibitive and require special care. All these add upto the costs and hence necessary investment due to sensitiveness of the area has to be taken into account. Eastern Ghat's bauxite deposits are mainly located in tribal areas and in some cases the terrain is highly dissected and rugged. The development of new mines in remote areas requires heavy investments in physical infrastructure. This may, however, be compensated by opening large mines to reduce unit investment cost.

The damage to the environment due to mining to bauxite had been witnessed in Amarkantak mining area. Both leading aluminium producers, namely Hindalco and Balco mined their requirements but disfigurement, water pollution, deforestation, destruction of natural habitat, flora and fauna, etc. have been reported. Damage control exercises have been taken up by the project authorities under the direct supervision of state and central anti-pollution agencies. Restoration work in mining area was also undertaken. The area has been declared as eco-sensitive area.

Marble

Production of marble is reported mainly from Rajasthan, U.P. and Gujarat and in small quantities from A.P., Bihar, Haryana and M.P. Production from Rajasthan accounts for over 88% of the total country's production. As per current trend, there is unlikely to be any mismatch between supply and demand of marble in the Ninth Five Year Plan. India has total recoverable reserves of marble to the tune of 100.9 MT (as on 1.4.90). It possesses vast deposits of very good quality and texture and variety of colours. Export in 1994-95 reached a figure of 73,253 tonnes.

The use of marble has created a problem of environmental pollution. The marble processing units, cutting and polishing units are facing problem of the accumulation of marble powder, an useless residue. This porous white powder is what people breathe in and live within the vicinity of marble mines. Another acute problem is that of water logging. Their disposal has also become a great problem to the mine owner, creating an environmental pollution. An experimental programme to find its utility in manufacturing of building blocks and tiles has been undertaken. Building blocks by mixing marble dust and black cotton soil and marble dust and red soils disintegrate immediately after cooling. The crushing strength of cement marble dust block was also found to be too low. The results of mosaic floor tiles tested for breaking load for dry and wet test and absorption indicate that a proportion of 1:1:1.5 and 1:1.5:1.5 of cement in marble dust and mosaic chips produce good tiles.

In view of the above, it can be concluded that marble dust is of little use either for manufacture of building blocks with clay or for cement marble dust block. More studies are necessary. The marble dust, however, can certainly be used for manufacture of mosaic tiles.

SUGGESTIONS

1. The environmental damage caused by small mines are no less than others when considered in totality. There is a need for a careful and detailed study at least in major mining centres, both of major and minor minerals, for evolving suitable measures for prudent environment management.

2. Generation of reliable environmental baseline data for small mines has been a major constraint in the preparation of EMP due to their limited resources. Such data could, therefore, be generated for a cluster of mines and regional/territorial EMP prepared for monitoring the effectiveness of control measures in those small mines. It is understood that the Ministry of Environment and Forest (MOEF) has taken steps in this regard and approved projects for preparation of such EIA and EMP in Gurgaon (Haryana) and Alwar (Rajasthan).

3. The unwary tradition bound workers, supervisors and negligent employers need to be made aware of the damages likely to be caused to health and life in general due to pollution of air, water, noise etc. and how to mitigate them. The mines safety weeks organised by the DGMS, and mines environment and mineral conservation weeks organised by the IBM held annually are steps to bring out awareness to the problems. It is suggested that more such studies should be carried out in various parts of the country.

4. The annual environmental awards instituted by the Federation of Indian Mining Industry (FIMI) for best performing mines has helped promotion of control measures in all mines. It is suggested that more such workshops on environment friendly mining should be organised in various parts of the country, jointly by IBM, FIMI, NISM and other NGOs with support from the concerned authorities in the Central and State Government.

5. The use of surface miners wherever possible need to be encouraged as drilling and blasting operations are eliminated resulting in cost reduction in operation as well as in avoiding noise and dust pollution. In this machine the ripping, crushing and loading operations

are combined.

6. Use of advanced computer-aided softwares for mine planning not only enable reliable estimation of mineral reserves, grade and effective blending and pit design but also facilitate environment friendly and cost effective mining through optimisation of production schedule.

7. Keeping in view the local conditions and other trade off between cost and benefits frequent dialogue between industry and pollution control authorities is necessary to voluntarily evolve an action plan and set targets for phased implementation.

8. Combination of in-pit crushing and conveying systems have become more acceptable in quarrying worldwide, since overall costs in comparison with truck haulage is less. Mobile crushing have high productivity, maximum utilization and high flexibility when used in conjunction with flexible belt conveyor systems and this would find more application in the industry.

4

Sustainable Agricultural Practices and Food Security

P.K. Shetty

While agriculture has been practised for the last 10,000-12,000 years, massive conversion of natural ecosystems into croplands has occurred during the last 150 years. According to the noted historian John Richards, between 1860 and 1920, 420 million hectares of land were converted worldwide to regular cropping, of which North America alone accounted for 164 million hectares, followed by Russia (88 million hectares) and Asian countries (84 million hectares). He further calculated that between 1920 and 1978, another 419 million hectares of forest and grassland were added to the world's cropland . The Indian subcontinent was not far behind in this expansion of land area used in agriculture. During the period from 1890 to 1970, more than 30 million hectares of vegetation cover was cleared for crop production and human settlement. During this period the amount of land under cultivation rose to over 45 percent and the population grew by 147 percent.[1] Even though farmers were expanding the area under cultivation, they were not able to provide the required amount of food for the growing population. In the post-war period, farmers adopted modern agricultural practices; as a consequence of which intensity of farming escalated.

During the last 50 years, scientific agriculture has enhanced world food production to remarkably high levels, from 700 million tonnes to approximately 2000 million tonnes per year. India, a land of varied climates and soils, providing a wide diversity in agriculture with about 100 million farming families has also experienced a similar trend. In 1950 the country's grain production was 52 million tonnes, and in 1998-99, it has risen approximately to 200 million tonnes.

The success of scientific agriculture while it supported the growing population has posed a big threat to the natural environment. The high intensity of farming has adversely affected the quality of land and water. Agro-chemicals have become a new source of pollution but increased food and livestock production have attracted many entrepreneurs into agro-based industries. Today the agro-ecosystem has to bear the burden of enormous amounts of industrial and agricultural waste products. There are several issues that need to be focused in the context of sustainable agriculture, for instance, erosion of natural soil, loss of biodiversity, salinization and indiscriminate use of agrochemicals. It is estimated that, the present-day world average of 0.28 hectares of cropland per capita is expected to decline to 0.17 hectares by the year 2025. It is very important, therefore to protect cropland and its soil and water resources for meeting the most basic needs of human kind.

The erosion of the soil is a natural and extremely slow process that has been accelerated by massive deforestation, soil tillage and over-grazing. The formation of an inch of top soil requires a minimum of 1000 years, depending on the climate, nature of parent soil materials and other factors. Top soil is more suitable for agriculture because it is rich in humus and plant nutrients. Microbial degradation of plant and animal residues as well as toxic agro-chemicals is high in this region. Soil erosion in the agro ecosystem through the action of wind or water affects land productivity as soil fertility, tilth and the water-holding capacity of land cannot be maintained. Each year about 7 million hectares of cropland are being lost through soil erosion particularly in dry tropical environment. According to Lester Brown, World Watch Institute, every year approximately 24,000 million tonnes of top soil are being eroded globally. This is equivalent to all the top soil on Australian wheat lands and represents the loss of 9 million tonnes of potential grain harvest. No doubt ecological imbalances caused by human activities brought about a measure of human suffering in history. For instance, the extinction of the Maya civilization of

Central America has been attributed to unchecked erosion and deforestation.

In India, it is estimated that between 11-26 percent of agricultural production is lost as a result of soil erosion. For instance, the annual soil loss is accounted at about 6000 million tonnes and which carries away 6 million tonnes of plant nutrients,[2] The croplands which have lost the natural source of plant nutrients cannot be restored only through the synthetic fertilisers. Moreover, food produced from such land may not carry the same nutritional quality. Besides, microorganisms are known to play a significant role in ecological cycles particularly in the degradation of toxic agro-chemicals, as a result of erosion this ecological process may be disturbed and many undegraded and partially degraded chemicals will poison the ecosystem. Wind erosion may disturb the agro-ecosystem by deposition of eroded materials on productive land. On the other hand, erosion will bring down the water quality, and also the life of reservoirs by progressive accumulation of silt over the years. Today, many reservoirs in India have the problem of over-silting.

To achieve the aim of sustainable and productive farm land, it is necessary to protect croplands from erosion by practising proper conservation methods. Simple practices like planting of hedges or rows of trees perpendicular to the direction of wind will reduce the wind erosion. Further, planting of cover crops during slack seasons or leaving the stubble in the fields following harvests will improve the soil stability and moisture retaining capacity. Erosion due to precipitation can be minimized by proper cropping practices such as mixed cropping, inter cropping, strip cropping and contour farming. Crop varieties which have spreading characteristics with an extensive root system are found to be more effective in reducing erosion. Farming practices across slopes, such as contour bunding and bench terraces will also enhance soil and water conservation.

During the last 150 years about 1000 million hectares of the world's natural or semi-natural vegetation cover was brought into cultivation. This process may have disturbed many associated plant and animal species which resulted in the reduction in the biodiversity of the agro-ecosystem. Further, adoption of monoculture by the farmers, and the use of high yielding varieties in place of traditional varieties have also led to significant loss of genetic diversity. Many of the lost plant and animal species may have carried away important genetic information with them. It is notable that 25 percent of the pharmaceuticals in use in the United States contain ingredients originally derived from plants. One of the biggest breakthroughs against cancer in recent decades has come from Madagascar Periwinkle (*Catharanthus roseus*) which is a source of two potent drugs used against Leukemia and Hodgkin's disease. The worldwide sale of plant-based drug items is now worth nearly US $30,000 million a year.[3]

Expansion of agricultural land by clearing of farm hedges is also a cause for concern because they harbour a high percentage of small mammals, reptiles, birds, butterflies and insect breeds. Farm hedges may provide some protection to the croplands by reducing the velocity of eroding agents and also harbour natural predators of agricultural pests.[4] Extinction of plant and animal species is a part of evolution but today due to human interference, the extinction rate is 1000 times higher than the natural rate. According to Peter Raven, a well known botanist, in the coming decades there will be a high rate of extinction at an average of 100 species a day.[5]

Salinization is a potential problem in the environment. If farm land becomes too salty, it will not support the crop growth. Every year 1.5 million hectares of cropland are being lost worldwide to salinization or water logging. In nature, soluble salts such as sodium, calcium, magnesium, potassium, chlorides, sulphates, carbonates, bicarbonates etc., are formed by the action of weathering of rocks or during the process of soil formation. Fertile land suffers from salinization either by improper leaching of salts or inundation by sea water. Some of the poorly drained perennial irrigation schemes are a big threat to the agro-ecosystem because they enhance the concentration and also make the land unproductive. India has approximately 7.5 million hectares of salt-affected soils and atleast 20 percent of irrigated land suffers from high salt content. One of the ways to reclaim the salt affected land is by the process of desalinization but this is expensive. However, some preventive measures, such as lining of irrigation channels, proper drainage of excess irrigation water to avoid salt accumulation, and intermittent ponding of water to have a better leaching of salts, have been found to be effective.[6]

Green plants require several mineral nutrient elements for their growth. At least one of the 24 essential elements is specific for a particular group or species of plant for proper physiological function. The widely used chemical fertilisers consist of three major essential elements that is nitrogen, phosphorus and potassium. The farmers were unaware of synthetic fertilisers until 1908 when the German Chemist Fritz Haber produced ammonia synthetically from nitrogen and hydrogen; using osmium as a catalyst. Further Haber's discovery led to the synthesis of urea on an industrial scale with the assistance of Karl Bosch. The discovery of nitrogenous fertiliser and its immediate acceptance by farmers resulted in the development of many new chemical fertilisers.

India since ages has been the user of environment friendly manures. Only after independence, the country has gone in for the massive production and use of nitrogenous, phosphatic and potash fertilisers. The new high-yielding crop varieties are most responsive to agro inputs. In order to get quick returns, farmers have followed monoculture instead of diversity in cropping. Farmers have moved away from their common practices of using organic manure and crop rotations with nitrogen fixing legume. The high yielding varieties and use of chemical fertilisers have no doubt increased cereal crops, namely wheat and rice in India, whose yields have increased by 50 percent and by 25 percent respectively. During the last three decades, fertiliser consumption has increased 60-fold. In 1960-61 Indian farmers used a total of 2.12 lakh tonnes of chemical fertilisers, in 1990-91 it rose up to 127.52 lakh tonnes.

India ranks 4th in the world in nitrogenous fertiliser consumption. Plants take up applied nitrogen in the form of nitrate or ammonia. Since, the 1960's there has been mounting concern about increased nitrate concentration of farm lands, especially the high level of nitrate in river and ground water. In Britain, it is reported that approximately 15 lakh people are exposed to nitrate levels in drinking water in excess of the limit set by the EEC standard of 50 mg per litre. The health risk from nitrate is often due to converted compounds such as nitrite or nitrosamine. If drinking water contains an excess of nitrate, it causes a fatal condition known as methaemoglobinaemia in young babies. In this case the nitrate in the human gut gets converted into nitrite by the action of bacteria, and the nitrite combines with the red blood pigment haemoglobin. As a result of this the oxygen carrying capacity of the blood cells is affected, resulting in suffocation. Apart from this, nitrite is a potential source of cancer, through the formation of N-nitrosamine.

The amount of nitrate in a particular ecosystem is dependent on the quantity of nitrate being added and cessation of plant uptake, precipitation, rate of leaching, type of soil, microbial activity and organic matter content. Nitrate levels in water build up slowly on the environment. Nitrate pollution in agro-ecosystem is often accelerated due to the addition of industrial or urban effluents. Continuous addition of plant nutrients through surface run off will increase the nutrient levels of an aquatic system. Nutrient enrichment of a fresh water lake or stream leads to rapid growth of algae and deoxygenation. As a consequence, many of the associated aerobic plants and animals are killed due to lack of oxygen. Due to increased use of synthetic fertilisers, the production and use of traditional manures has been neglected. It is essential to look into strengthening this low investment oriented sector because the country has the enormous amount of raw materials for manufacture of environment friendly manures and also which will help to promote local employment.

In the early 1960's, Rachel Carson's famous book 'Silent Spring' exposed the catastrophic impact of pesticides on a fragile environment. Since then, an enormous amount of research has been conducted which has shown the presence of pesticides and their degradative products in soil, air, water and on living systems. Pesticides are toxic chemicals intended for prevention, destruction, mitigation or repelling of pests. Pesticides are commonly used to kill crop pests, plant pathogens and weeds. The use of chemicals has also been extended in the storage of foodgrains, public health programmes, and in households sprays etc.

Annual global pesticide consumption has been increasing progressively. It is estimated to be approximately 50 lakh tonnes. Over-dependence and improper use of pesticides has led to pollution, biomagnification, and destruction of non-target species, including beneficial natural enemies of pests. Though pesticides have provided a temporary solution to pest problems, they have long-lasting effects on the ecosystem. There have been many occasions when pesticides have caused serious damage. Since 1933 there have been about 100 accidental outbreaks of acute pesticide poisoning, many of them related to contamination of food or water, and occupational exposure.

The World Health Organization and the United Nations Environmental Programme, in 1989, reported that pesticides cause about one million cases of accidental human poisoning and approximately 20,000 deaths each year. It is interesting to note that developed countries account for 80 percent of the total global pesticide consumption but record less than one percent of pesticide-induced human deaths. Several survey reports have indicated the presence of chemical residues in human food, dairy products, fruits and vegetables, animal feeds and even breast milk. It is true that pesticides have become an essential part of modern agriculture and have played a significant role in increased food production. However, excessive pesticide use has not completely eliminated crop pests, and still many countries in the world continue to suffer from crop losses. For example in the United States about 37 percent of food and fibre crops are lost each year to pests. The loss of agricultural produce to pests and diseases in India amount to some Rs. 33,660 crores a year.[7]

As the threat to the environment from toxic chemicals continues, there is need for increased public awareness of the facts about these chemicals. In this context, it is relevant to mention that many chemicals which are banned in the West are still being used in developing countries like India. Governments should regulate the use of the banned pesticides, and should encourage environmentally safer pesticides. Most pesticide related poisoning and deaths in the developing world have resulted from inadequate occupational and safety standards, the user's unawareness about the toxicity of the chemicals and the proper handling techniques. It is extremely important to educate farmers on need-based and judicious use of toxic chemicals and also the concept of integrated pest management practice should be promoted to encourage sensible use of chemicals.

Finally, we must stress the value of new thinking and practices for Indian agriculture. Over 80 million out of the 90 million operational farm holdings in India are below 2 hectares in size. It is essential to assist the small farmers to improve the productivity and profitability in their farms. Today, the Indian farmers have to provide food for some 980 million people and fodder for over 400 million livestocks. Apart from this, fuel wood is a source of energy for a large population. In future, a sustainable farming system needs to be created through a proper mix of both non-conventional agricultural practices and modern technologies which are environmentally safe and profitable, for producing more food, fodder and fuel wood through the optimum use of the land and water resources.

Notes and References

1. Hrabovszky, J.P. (1985). Agriculture: the land base. In: The Global possible resources, development and the new century (ed.) Robert Repetto. Affiliated East-West Press Pvt. Ltd., New Delhi, pp. 211-254.
2. Malcolm S. Adiseshiah (1987). Economics of environment. Lancer international in association with India International Centre, New Delhi.
3. Norman Myers (1991). The Disappearing Forests. In: Save the earth (ed.)

Jonathon Porrot. Dorling Kindersley, London, pp. 47-50.

4. Joy Tivy (1990). Agriculture ecology. Published by Longman scientific and technical in association with John Wiley and Sons, Inc., New York.
5. Endangered earth, in Time Magazine, January 2, 1989.
6. Biswas, T.D. and Mukherjee, S.K. (1987). Soil Science, Tata McGraw Hill Publishing Co. Ltd., New Delhi.
7. Dhaliwal, G.S. and Arora, R. (1996). An estimate of yield losses due to insect pests in Indian agriculture. Indian J. Ecol. 23(1): 70-73.

5

Urban Ecology

NABA KUMAR PATNAIK

"Usually in rural areas, functions are mostly traditional but in urban centres they are more diversified. It is quite obvious that rural occupation is dominated by agriculture and its associated activities; while in urban centres the main activities are manufacturing, trade, commerce, transport, communication and many professional, personal, official and institutional services. Thus the secondary, tertiary and quaternary activities are the characteristics of urban occupational structure." (Maurya, 1982).

Modern Indian cities are attributed to rural-urban migrations. The migrants who pour into urban centres in search of jobs and opportunities are in really 'ecological refugees', who can no longer sustain themselves in the highly vulnerable rural environment and who despite the inhuman conditions that await them in urban slums choose to migrate to them due to relatively better accessibility to jobs, health services and education. However, the major problem of the Indian cities is not just immigration of the poor and the needy who form hardly 10-20% of the urban population. While the majority comprising of the very rich and middle class have a very high rate of consumption of resources such as water, electricity, fuel and other commodities, the access of the poor to these resources itself is limited by lack of infrastructure. Most of the pollution too inherent in the cities is caused by industries, vehicular traffic, strewing of garbage and misuse and abuse of water and sewage systems such as illegal or unlimited tapping of water lines, bore-wells and even illegal connections between the sewer and storm water drainage lines. Hotels, hostels and Kalyan Mandapams are also the largest consumers of fuel wood in large cities. Besides, house building and brick making are also causes of deforestation in rural areas of forest and timber species. Since urbanisation, by its very definition, relies on specializations, people expect certain basic amenities to them from the Government and other private institutions. This makes the people dependent and helpless.

HEALTH

The health of a nation is not dependent on the number of doctors and hospitals. For instance, the most important environmental factor linked with health or rather ill-health is water. Eighty percent of world's diseases are linked to water. As an indicator of health, the number of water taps per head of population is more important than the number of hospitals or doctors, says WHO's Director General, Halfdan Mahler.

There are 5,100 hospitals situated mainly in urban areas, with a number of them providing the most uptodate facilities for the 109 million urban dwellers. Several reviews of health services in the country have pointed out the urgent need for primary health care programmes. An integrated approach to health care would also mean preventive programmes that attack the environmental factors responsible for the spread of disease. Lack of such an integrated approach presages a menacing future for the health of country.

- With the sole exception of Smallpox, almost all important diseases like tuberculosis, leprosy and diarroeas continue to be rampant, and there are no immediate prospects of these diseases being reduced drastically or eradicated.
- Malaria and Kala-azar thought to have been eradicated have returned.
- Diseases like Hepatitis and Filariasis and many new and killer diseases like Dengue and Japanese encephalities are spreading because of the changing rural and urban environment.
- Commercial pressures and so-called modernisation trend

are combining to bring in a range of diseases—which can be termed commer crogenic diseases—like cancer, exacerbated by the sales pressure of the tobacco industry; infant charrhoeas leading to death, due to sale pressure of the milk powder industry; and anti-biotic resistant strains of diseases caused by the combined pressure of the medical profession and the drug industry to over-prescribe drugs.

Water and Infection

There are certain trends within the water supply and sanitation sector which are disturbing from a public health point of view.

First, even though the proportion of the urban population covered by Municipal water supplies is increasing—it now stands at over 80 percent—the quality of water supplies has been constantly deteriorating. People in many of the largest cities of India who received running water for 8-10 hours daily in 1970, received only 2-3 hours in 1975. Nothing is more dangerous to public health than intermittent water supply. The chances of sewage water from leaking sewer entering the nearby water pipes are high when pressure in the empty water pipes is less. An intermittent supply of water leads to a negative pressure within the water pipes, drawing in sewage and faecal material. The contamination of drinking water is seldom at the source.

As such as 90-100 percent of the urban population in the states of Gujarat, Maharastra, Manipur, Nagaland, Orissa, Punjab, Rajasthan, Tamilnadu and West-Bengal is receiving intermittent water supplies.

It is no wonder that the incidence of hepatitis (Jaundice) has been increasing dramatically in most Indian cities—from there it is now slowly spreading to rural areas. Its incidence in the last five years may be more than that of the previous 15 years combined.

Second, the unfortunate trend in the public health system is the growing gap between water supply and sanitation services. Although 2,092 urban communities have been provided with piped water supplies, only about 217 towns provided sewage system out of 3,126. In rural areas, the gap is even bigger.

The fundamental health reason for supplying clean drinking water to reduce the incidence of water borne diseases. But increasingly, scientific evidence shows that supplying clean drinking water is not enough.

Many skin and eye diseases (Seabies, skin sepsis, fungal infections and Trichomonos) are not water borne at all but they are greatly influenced by water use practices in a community. Using street taps had a higher incidence of diarrhoeas and a bacillary dysentery which can hit children badly. There is a high risk of water contamination during transport and storage. Clean water was often collected indiray containers. There are other infections that are spread by fleas, ticks, lice and mites. All these diseases can be classified as water-washed diseases and their incidence is likely to be affected more by the quantity of water than by its quality.

These clearly show that drinking water supply programmes must be accompanied with sanitation programmes aimed at improving personal hygiene, if public health benefits possible from community water supply schemes are to be obtained. Unfortunately, in India today, there is little interest in pushing sanitation and health education programme.

In urban areas, slow progress in sanitation programmes is due to technological perceptions—the undue emphasis on Sewage Systems to the total exclusion of cheaper non-sewerage alternatives.

Sanitation Essential

Moreover, dirty hands transmit enough germs to keep the incidence of diarrhoeas within a community high even if it has a community water tap. Water consumption from community taps usually remains limited to drinking and cooking purpose. The total quantity consumed varies according to the distance from the tap to the house. There is inadequate water to keep washing hands all the time in most poor households. If the nearest tap is 1,000 meters away (that is in real situations if a poor family's lucky) hardly any one will be committed enough to cleanliness to traverse that distance dozens of times each day.

A survey by the World Bank found that the highest diarrhoea infection rates were in households which are furthest away from their water sources. Those families with taps inside the house tend to have the lowest infection rates; those with water close to the house have the next lowest. Infants and small children, who generally have a high rate of diarrhoeas can get infected frequently and repeatedly not by drinking unsafe water, but by Parents who do not practice good personal hygiene.

Development and Diseases

Equally worrying is the way economic development activities are contributing to the spread of disease. Irrigation Systems, for instance are a major cause of the increase in the incidence of Japanese encephalities. Similarly, the construction of dams in southern India in areas where well-water already has high quantities of flourides, has changed dramatically the character of the disease called fluorosis. It now affects even the young and makes them severely knock-kneed.

In many towns, water supply schemes have been built without provision for proper drainage. The consequent stinking of waste water becomes excellent breeding grounds for mosquitoes. Open-drains are quite common sight in many villages and towns.

Unplanned urbanisation, construction of water projects, migration of people—all help to change the ecology and epidemiology of diseases. As V. Ramalingaswami, Director General of the Indian Council of Medical Research points out in no uncertain terms. "The health implications of development activities should be given serious consideration before the activities are commenced". He says that activities which have an impact on the environment and ecology need special attention.

Smoking and Cancer

The most worrying phenomenon is increasing use of bidis and cigarattes in India which would result in a dramatic surge in lung, cancer. According to the WHO, smoking kills atleast one million men and women every year in world. 90% of all lung cancer deaths, 25% of deaths from cardio-vascular diseases and 15% of deaths due to chronic bronchitis are directly caused by smoking. The close association between the marketing practice of the tobacco industry and the increase in smoking qualifies for human-made "carcingenic" disease.

The Government has done little to control or reduce smoking because of conflicting loyalities, the needs of the exchequer and the health of the people. India is the third largest producer of tobacco in the world after the US and China. Apart from earning foreign exchange, tobacco and tobacco products provide a consistently increasing source of revenue. The tobacco industry also generates vast employment. Bidi smoking is a major cottage industry in India and provides jobs to millions of people and in large number of cases, it is the sole means of their livelihood.

In recent years, the Government has been trying to take the pressure off the cigarette industry by not increasing excise taxes any further. Consequently, cigarette sales increased by six percent. Having experienced an average annual growth of barely two percent in the last several years, the tobacco industry is now planning for faster growth unhampered by new taxes.

Meanwhile the Government's efforts to introduce statutory warnings on cigarette packets and advertisements about the health hazardou of smoking are neither being enforced properly nor are they having an impact on smoking habit. Several leading health experts fear that even before malnutrition and communicable diseases are brought under control, developing countries like India could be overtaken by the spreading of cigarette smoking habit. Strong measures are required in India to control the looming smoking induced cancer epidemic. Glossy bill boards depicting young men and women enjoying a casual puff while leaning on a car are to say the least, a cruel social joke in a country with a per capita income amongst the lowest in the world.

HAZARDOUS PRODUCTS

The world to-day produces chemicals faster than it can manage them. Nearly five million chemicals have been synthesised in the last 40 years at a rate of 10,000 new ones every month. Some 50,000 to 70,000 chemicals are used extensively in millions of different commercial products.

These chemicals include extremely toxic substances which can cause allergies, damage vital organs of the human body like the eye, brain, liver, kidney and reproductive organs, produce malformations in unborn children and even in generations to come, and cause or promote cancer, in quantitits as small as parts per billion, equivalent to a few drops in an Olympic-size swimming pool. In case of accidental release into the environment in large quantities—as happened in the case of methyl isocynate (MIC) in Bhopal—they can lead to mass mortality. What is amazing is that the world knows so little about all the chemicals it uses. A recent US study indicated that no toxicological information whatsoever is available for about 80 percent of all Chemicals used in commerce. Not surprisingly, these chemicals keep throwing up surprises with their deadly properties.

Such chemicals which are now an inevitable part of the industrialisation process pose hazards to human health and environment in three different ways. Firstly they expose workers—industrial workers and increasingly agricultural and plantation workers—run high health risks. Secondly industrial wastes and effluents and run off from agriculture, steadily contaminate the environment and pose long-term risks to human health and the life of other biota. Occasionally accidental release leading to catastrophic pollution, can lead to mass deaths. Thirdly use of chemicals in products of daily life—as artificial sweeteners, flavouring and colouring agents, extenders of shelf-life in processed and packed foods from milk to tomato sauce, as paints, detergents, cosmetics and manufactured drugs and as various forms of plastics and fibres—they pose an insidious threat to human health. For example, lead, a dangerous metal, today exists all around a human being in water pipes, in pesticides, in industries which involve lead smelting and refining, storage batteries, lead based paint, ceramic crockery, containers with lead glaze copperware lined with impure tin containing led in newspaper and magazine ink. People in cities are exposed to lead in car exhaust fume. Those living around steel mills are exposed to lead in air—425 million tonne (mt) capacity plant means 80 kg of lead setting on every square kilometer in the vicinity of the plant.

There are number of separate acts which deal with aspects of the chemical safety. But even when put together, they are totally inadequate to deal with the emerging safety problems, which arise at all stages of chemical manufacture, import, storage, transport and handling and finally even at the points of sale and use in agriculture and industry and by the general consumer.

The number of institutions which have properly trained human power and have adequate facilities to determine the safety of chemicals is only three, the National Institute of Occupational Health at Ahmedabad, the Industrial Toxicological Research Centre in Lucknow and the Central Labour Institute in Bombay. In recent years, the Government has set up a chain of five labour Institutes across the country to deal with problems of chemical safety.

Today unfortunately, urban ecology is no more sound and is posed with health hazards and impaired human activity, due to low per capita availability of land.

Notes and References

1. Balasubramaniam, Arun, Ecodevelopment: Towards a Philosophy of Environmental Education—Regional Institute of Higher Education and Development, Singapore, 1984.
2. Futehally, Laeeq, Our Environment, National Book Trust, New Delhi, 1992.
3. Krishna, Shyamala and Dange, Archana, Addressing Urban Environmental Issues through Environmental Education—Environmental Education for Sustainable Development, Indian Environmental Society, New Delhi, 1994; pp. 81-84.
4. Mishra, Ranganatha, Concepts of Environmental Studies, Pustak Mandir, Berhampur (Gm.), 1996.
5. Centre for Science and Environment—The First Citizens Report, New Delhi, 1982.
6. Centre for Science and Environmental—The Second Citizen's Report, New Delhi, 1985.
7. Indian Institute of Ecology and Environment—International Encyclopaedia of Ecology and Environment, New Delhi, 1994.
8. Indira Gandhi National Open University, Human Activities and Environment, New Delhi, 1993.

6

Bioethical Issues in Livestock Industry

M.V.G. Ahobala Rao

Animal bioethics is a constitution of integrated ethical principles aimed at not only making the animal's life comfortable, productive, painless but also environmentally sustainable for the present and future generations. It also should aim at the protection of mankind from the ill effects of the sick and unproductive animals. Animal is the dominant partner of our ecosystem deserving due attention for its existence, productivity and growth. Earth Summit—1992—agenda-21 had endorsed a programme where in the participant countries have committed to a series of practical and legislative steps for the protection and conservation of natural resources of which animal is an integral part. Emphasis was made on the environmental accounting and audit—in this agenda, sustainable livestock production should be aimed at the recognition of the fact that natural resource base is finite and must be preserved, extracting most from the least without inflicting permanent damage to it, taking due care on the health of man and animals and protecting them against pollution, infiltration and contamination, supporting the poor and ignorant in protecting the environment, making every one concerned to be involved in the process of decision-making on the steps to be adopted for environmental protection ensuring adequate and safe feed, water shelter and health care to every animal and finally updating the environmental friendly technology in the livestock maintenance.

Erosion of traditional social systems is sweeping away the entire welfare and security nets of traditional communities, creating vulnerability and dependence, resulting in increasing urbanisation and consequent ecological imbalances, foodinsecurity, hunger and poverty coupled with increasing population are the major threats facing our ecosystem. Forest fires, volcanos, oil spill overs, nuclear hazards, floods and cyclones, chemical pollution of water, soil and air are endangering the life on the earth.

Since, animal is an integral and most important part of out ecosystem, we have to understand its impact on the society and the environment and vice versa. It is note worthy that about 2/3 of world's livestock is in the developing nations and Indian live stock population constitutes about 1/5 of the world's livestock population, it is happy to mention here that India has already attained first position in the world in milk production replacing U.S.A. recently. India is the fifth largest egg producer in the world producing about 30 billion eggs every year.

Livestock Population

(in Millions)

Kind of livestock	*World*	*India*
Buffaloes	140	75.0
Cattle	1279	193.0
Sheep	1190	54.5
Goat	557	110.0
Pig	857	140.0
Poultry (Layer) annual output	—	152 million
Poultry (Broiler) annual output	—	830 million

Contribution of livestock to the Indian rural economy is Enormous. In Kaira District of Gujarat state, where the world famous diary cooperative AMUL is operating, income from dairying constitutes over 47% of total farm income. 76% of milk produced in India is from the tiny small farmers and 90% sheep are owned by poor farmers and almost 100% piggery is in the hands of peasants. 40% of eggs produced in India is from weaker sections. These figures show the impact of livestock in the socio-economic development of a nation where over 75% of population thrive on rural farming. Socio-economic freedom and development of rural women in India was achieved during past few years, mostly through livestock only as women are the main participants in livestock husbandry which fetch them significant and almost stable employment and income round the year unlike from agriculture sector.

Man's dependence on animal is multifaceted. Animal is food factory supplying milk and meat. It is a powerhouse, as about 2/3 of farm power requirement (in the form of drought energy), is from work animals providing employment over 20 million people saving fuel worth about Rs. 45,000 crores annually. In India over 14 million bullock drawn carts are on the roads, Bullock is still considered as the backbone for Indian agriculture and rural transport since prehistoric ages. Animal is considered as fertilizer factory providing enormous quantitites of manure to enrich the fertility and texture of the soil in a most eco friendly manner.

Bioethics in livestock industry is concerned with our role and responsibilities and consequences of our life styles, our attitudes and values, policies, activities, processes, products, in much wider sense than the other ethical systems. It should be remembered by every one that equality of consideration is the basic right of every animal. Dr. Adrew Fraser was the first veterinarian to recognise the importance of bioethics in veterinary practice and education. Love and kindness towards dumb animals is the inherent character of ancient Indian culture, where Cow, Bullock, Elephant, Tiger, Lion, Monkey and even Cobra and a number of other animals are considered very sacred.

Bioethics is considered as a synthesis on ethics, scientific and empathic knowledge in the field of ecology, ethology, cultural anthropology, and social economics. An ethical approach to veterinary education and practice can play a very significant role in helping the global society to become more humane, compassionate, economically and spiritually viable and ecologically sustainable. The study and practice of bioethics should be made a part of veterinary profession and even in the primary education itself.

Bioethical issues in livestock industry may be classified and studied in the following manner.

1. ZOONOSIS CONTROL

Many fatal diseases like tuberculosis, anthrax, Brucellosis, Pneumonia, Q-fever, leptospirosis, sarcocystosis and many non-specific infectious diseases of animals are communicable to man through direct contact, feed, water and inhalation and also through indirect contact. Bovine spongiform encephalitis popularly heard as Mad Cow disease in recent years had made Great Briton virtually mad by demanding for mass destruction of large herds of valuable cow population in 1986. Similarly the recent out break of Avian Flue in Hong Kong in 1997 caused by HS-NI Virus into man, posed a great threat to that nation resulting in mass destruction of millions of birds in the public health interest. Rabies is the second most fatal viral disease to man next to AIDS. Children get skin diseases like dermatitis mange, scabies etc., from their affected and carrier pets apart from several gastrointestinal parasitic infections of zoonotic significance.

2. FOOD HYGIENE

The quality of feed and water taken by the milch and food animals has direct bearing not only on the health of the animal but also on the consumers of their milk and meat. Chemical pesticides, weedicides, antibiotics, growth promoting drugs used in the treatment of sick animals and chemically polluted water and feed resources show disastrous effects on the health and long term productivity of animals. People thriving on such animals also suffer from toxic effects. Hence, the livestock should be kept in a clean and healthy environment, where the animal does not have any access to contaminated air, water, feed and soil. For attaining this a stringent legislation prohibiting the release of un treated toxic wastes into the natural resources to be enforced. Slaughter houses and dairy plants should have modern affluent treatment plants and adopt hygienic processing, packing, storage and transport facilities with aim on physical, chemical, bacteriological and aesthetic quality of the final products.

Control of Environmental Pollution

It should be remembered that the environment and the animal are both polluting each other. Animals are afflicted with air borne, and water borne diseases from the unhealthy environment, automobile fumes, industrial exhausts and affluents getting access to the animals with residues of toxic chemicals like Lead, Zinc, Chromium, Cadmium, Arsenic, Mercury, Fluorine, Dioxides, Chlorofluorocarbons, causing serious health hazards in animals. Methane production by ruminents in enormous quantities is of serious concern. An estimated 13.2% of World's methane is liberated by the ruminating animals. Methane constitute 18% of total green house gases that destroy Ozone layer. Methane is 40 times more effective than carbondioxide, is destroying Ozone. Ruminents thriving on cheap quality roughages like paddy straw liberate more methane. Scientific feeding with succulent fodders, balanced concentrate feeds, Enriched straws with urea and Ammonia and addition of antimethanogens like Rumensins and monensins in the feeds, maintenance of rich grazing fields, can substantially reduce methane production by ruminents. Incidentally, these same practices are suggested for better health and better productivity of animals also. Reducing the size of cattle herds by timely culling and slaughter of unprofitable animals is also very important recommendation in this regard. Ingestion of sharp Metallic objects, presence of bio-un-degradable plastic and synthetic material in the garbage that attract hungry animals are posing serious threats to their lives after causing a lot of miserable suffering mainly in the urban areas. Untreated hospital wastes in the garbage places is endangering the health of both animal and man. Scientific and ethical disposal of industrial and domestic wastes has to be given due importance. Drought animals on the urban roads are exposed to serious chemical pollution from automobile exhausts, that lower their health, productivity and also reproductive efficiency.

Research on Comparative Biomedicine

There are many hazard free and effective herbal remedies available today for the treatment of animal and man. They are cheap and cause very little side effects on the patients. Allopathic treatment with synthetic chemicals many a time lack long term positive effect on the patient. Many herbal products have multifaceted actions including immuno modulatory effect to trigger the body defence mechanism to counter the effect of many disease causing agents. The herbs are eco friendly. Hence, research to identify improve, and conserve our herbal wealth and their processing and storage and use need special attention. A global knowledge bank on herbal medicine through computer network can go a long way in the world wide use of herbal remedies both for effective and economical treatment of sick. Fund allocation on these research projects be made with due priority. The forests, the treasure houses of our herbal wealth must be protected from forest fires, floods, and erosions and other destructive factors.

Improvement of Diagnostic Techniques, Drugs, Chemicals and Biological Products

Preventive medicine with improved sanitation, timely vaccinations, and supporting natural defence mechanism, improving the nutritional status of animals, improving diagnostic techniques and providing timely and accurate therapy to the sick go a long way to improve the over all quality of life of animals and their productivity and consequent profitability. Appropriate steps to isolate the sick and to maintain new animals under adequate quarantine have to be followed to safeguard the cattle and human health.

Regulation of Import and Export of Livestock and Livestock Products

Unchecked movement of livestock and their products may lead to serious health hazards both in man and animal. Hides, leather, wool, hair, meat, gut and other animal products may carry pathogens causing diseases like Anthrax, Tuberculosis, Tetanus etc. Stringent quality checks with well equipped laboratories are to be enforced in every port through which passage of animal and animal products should be insisted. With the faster and long distance transport facility available for animals and animal products, the scope for the spreading of infectious diseases in much wider areas also has enormously increased during past few decades, nececiating for more stringent quality control measures on the importing livestock and livestock products including animal dung as manure.

HEALTH EDUCATION

Educating every one concerned directly and indirectly with the animals and their products and services including attendants, traders, handlers, health workers on animal health, zoonotic diseases and scientific and hygienic husbandry practices, care of sick animals, and preventive health management needs special emphasis. Even children and old need to be educated through meetings, Press, TV channels, films, radio and other means of communication. It should be remembered that health education is a continuous process requiring intermittent re-orientation and follow up.

Emergency Care of the Sick

Immediate alleviation of pain and suffering is the basic ethical requirement of not only a veterrnarien, but of every human being. Relief from strain, provision of comfortable shelter, cool and clean water and food and providing necessary first aid and special care and taking the animal to the clinical veterinary care in time, are expected bio ethical services.

Humane, and Kind Attitude

Love towards any animal from the childhood to be promoted through appropriate teaching methods. Cruel attitude towards animals by the children has to be curbed at their budding stage itself. Kindness towards animal has many fold positive effects on the mental and physical health of man and his feeling of security, using handling and transporting of animals for various purposes should be carried out taking all possible measures to protect them from unnecessary strain discomfort and agony. Better and scientific training of work animals in the initial stages reduces a lot of physical burden on the attendant in the long run apart from increasing the life time work efficiency of his animal. Animal races where they are exposed to cruelty should be prohibited. Provision of suitable yokes, harnesses, limiting the load and provision of rest breaks after every prescribed stretch of work, using the work animals only during specified times so as to protect them from extreme climatic conditions, proper training to the drivers and attendants, provision of specialised shoeing to suit the work and road conditions, granting leave to the sick, pregnant and disabled animals are some recommended humane animal husbandry practices. At the point of extreme agony and suffering and where chances for the survival are very meagre and where the life of an animal poses serious threat to the security and health of other animals and public life in the vicinity, mercy killing is suggested with appropriate means of euthanasia resulting in painless death. Exposing animals to horrible pain, suffering and death during certain social and religious occasion in the name of animal sacrifice, is a shame on the man kind. Every abattoir should be equipped with stunning devices to ensure painless slaughter. Slaughter without stunning should be prohibited legally.

Teasing the work animals with cruel acts like housing in inadequate hazardous, and crowdy cages, boxes and rooms, depriving minimum scope for movement over crowding during transport in vehicles depriving food and water beyond reasonable time etc. are to be treated as serious acts of offence by appropriate laws and they should be enforced by an agency having representatives from police, legal and veterinary profession and animal welfare associations.

Social Concern

Existence of animals should never be an undue burden on the society. Competition by the animal for food and shelter, with man should be avoided as far as possible in the larger interest of both man and animal. Use of land and feeds that are not much required by man, can well be allotted to the animal. Non-conventional feed ingredients like many agro-industrial by products which do not affect man can safely, economically and profitably be fed to the animals. Inter species competition for food also can be avoided to a great extent by substituting the use of certain feed ingredients in the rations of some species which are not so essential for them but more essential for other species. Similarly avoidance of animal keeping in the urban areas also need special emphasis in view of several social and economic factors and for the security, of animal and public health. Economic maintenance without ignoring the well being of the animal, is also a vital issue aimed to make the availability of milk, meat, egg and other products at a price which is within the financial reach of even poor consumer. Careful, monitoring of the size of the livestock population to suit the available feed and water resources is also a matter of serious concern.

GENETIC MANOEUVERING

Creating Transgenic animals, introduction of incompatible genes, breeding for high productivity without any concern for their disease resistance and adaptability to the local climatic and nutritional stresses, use of bio-technology to create animals for the purpose of having them as organ donors for man are some of the important issues concerned with the genetic manouvering wherein bio-ethics has to play significant role.

Bio-ethics should be made a compulsory topic of study and practice in Veterinary medicine and a seperate cell need to be established in the state and national veterinary councils, to help the legislative, administrative and enforcing machinery in the matters related to bio-ethics in livestock industry.

Finally, when man expresses positive attitude towards his animal and when the animal is not afraid of man, and when there is better social bond between man and animal, sows will have more piglets, hens lay more eggs, work animal show better performance, cows produce more milk, meat animals produce more meat with better quality and pets show more affection and faithfulness, thus fulfilling the universal motto "Health and Wealth for All".

References

M.V.G. Ahobala Rao (1997), Animal and Environment.

———, Ozone and the Ruminant.

———, Pollution *v.* Dairy Industry.

Alexander Stephens (1998), Rural Sociology. . . Constraints.

R.W. Bell (1998), Bovine Spongifrom Encephalitis.

H. Chennegowda (1998), Milk Production at village level.

S.K. Dwivedi (1997), Effect of environmental pollution on livestock health.

Michael W. Fox (1998), Veterinary Bio-ethics for the common health and good.

M.A. Haleem & M.S. Rizwana (1998), Food processing . . and Technology in India.

C. Krishna Rao (1998), Livestock production at village level.

Lalkrishna (1998), Control of livestock diseases in India.

C. Natarajan (1998), Public health and the veterinarian.

N.S Ramaswamy (1998), Drought animal power role of Veterinary profession.

M.J. Saxena (1998), Therapeutic efficacy of herbs and shrubs in veterinary Medicine.

G.P. Singh and Madhu Mohini (1995), Methanogenesis in Ruminents and its effect on global warming.

G.P. Singh and Pat. Zimmerman (1997), A new method for estimation of methane from Tracer technique.

Lord Soulsby (1998), Veterinary public health—An expanded role for veterinarian.

7

Diversity of Terrestrial Algae and Cyanobacteria and their Ecological Significance

D.P. SATAPATHY AND S.P. ADHIKARY

Algae, although generally occur in fresh water or marine habitats, also occupy a variety of terrestrial environments where their ecological importance is considerable. A number of terrestrial niches are occupied by algae: most commonly, they occur either on the surface or at a depth of up to several centimetres in soil; live in and on rocks and in caves; inhabit on permanent snow and rice-fields and can also be found living on animals and plants. Most of these terrestrial algal habitats represent extreme environments characterized by aridity, and/or by low or high levels of temperature or light intensity. In addition, in many of these habitats, frequent fluctuations of environmental conditions occur in sharp contrast with the mostly stable aquatic habitats. To cope with these stress situations, terrestrial algae develop morphological and physiological adaptations and/or occupy sheltered microhabitats in which conditions are less severe. Among the different algal groups, the prokaryotic blue-green algae (cyanobacteria) were generally the major constituents of terrestrial algal populations.

Soil algae: The soil was the best studied terrestrial algal habitat. The most important environmental factors which control algal populations in soil seem to be light, humidity, temperature, availability of nutrients and pH. As most of the blue-green algae (cyanobacteria) were obligate photoautotrophs, the role of light was evident from their vertical distribution: the density of these organisms were generally highest in the upper centimetres and falls off very rapidly with depth (Peterson, 1935). Algae were often present far below the levels of light penetration, yet it was questionable whether these organisms were actively metabolizing or were in a dormant state. Chemoheterotrophy in the dark had been reported in many blue-green algae (Stanier, 1973; Khoja and Whitton, 1975 and Bastia *et al.*, 1993). These organisms thus could exist without the presence of light; it was, however, questionable whether they could effectively compete with obligate heterotrophs for the limited supply of available organic matter. The fact that generally no algal species were found in deeper soil layers which were not present at the surface of the same soil was a further argument against the existence of an independent subterranean algal flora (Petersen, 1935). The occurrence of the algae in such depths may be due to water seepage or to the borrowing activities of animals. In general, it seems that actively growing alegal populations exist in the top layer of the soil and that heterotrophy was of minor importance in soil algal productivity.

Most soils were subjected to fluctuations in the levels of humidity and soil algae were generally adapted to survive periods of desiccation. Viable soil algae (e.g. *Nostoc commune* Vauch., *Chlorococcum humicola* (Nag).) Rabenh., *Stichococcus bacillaris* Nag.) had been recovered from up to 87 year old herbarium specimens (Lipman, 1941). Seasonal changes in soil algal vegetation were generally quantitative, due to fluctuations in the availability of water, while the species composition remains constant over the year (Metting, 1931). Some desiccation-resistant soil algae from resting spores (e.g., *Cylindrospermum., Anabaena, Nostoc, Nodularia*) while certain green algae tolerate desiccation without apparent special morphological adaptations (e.g., *Prasiola crispa* (Lightf.) Menegh., *Hormidium flaccidum* (Kutz.) A. Br.).

The effect of pH on the algal flora was difficult to evaluate as it was often correlated with other factors. Thus, arid soils were almost universally alkaline and many continuously wet soils were acidic (Shields and Durrell, 1964). The pH also depends on the calcium carbonate content of the soil. Among the different algal groups, cyanobacteria (blue-green algae) generally prefer neutral and alkaline soils while green algae seem to tolerate a wider range of pH, so that they dominate the algal floras of acidic habitats due to the absence of competition. Cyanobacteria, especially filamentous forms, were common in neutral to alkaline soils. Species of *Microcoleus, Schizothrix* and *Porphyrosiphon* often form crusts on the soils (Durrell and Shields, 1961). Heterocystous diazotrophic genera such as *Nostoc, Anabaena, Tolypothrix, Scytonema* and *Cylinrospermum* were to great significance to the soil ecosystem. The role of algae in the soil ecosystem was manifold, the most important consequences being the input of nitrogen by diazotrophic cyanobacteria and of carbon. The formation of algal crusts on soil interferes with soil erosion, increases the storage of rain water and reduces the water loss by evaporation (Booth, 1941). They also provide organic matter and were responsible for soil formation which was especially important in deserts.

Lithophytic algae live on or within rock substrates, expanding to a few mm below the rock surface. Lithophytic algae generally colonize bare rock surfaces which were not covered by lichens or mosses. There were characteristic of extreme climates such as hot and cold deserts and in temperate climates they occur in extreme microhabitats like steep rock faces in the high alpine regions. Algae can compete successfully with lichens and mosses and were often the only autotrophic organisms colonizing the rocks. In response to the extreme ecological conditions and frequent changes in the environment, lithophytic algae were adapted to switching their metabolism off and on; metabolic activity was limited to short periods when an appropriate combination of temperature, light intensity, and humidity was present. This period of metabolic activity may be as short as a few hundred hours in a year e.g., in the Antarctic cold desert (Friedmann *et al*., 1987). In extreme arid environments only endolithic and hypolithic forms could exist. The microclimatic conditions seem to be more favourable inside and under the rock than at the surface where frequent temperature fluctuations (McKay and Friedmann, 1985), aridity, light intensity, and wind may prevent the establishment of an algal vegetation.

The most important factor concerning the ecology of lithophytic algae, especially in deserts, is water supply. In temperate regions, rocks inhabiting algae live in temporary water seepages and/or were regularly wetted by rain water. In deserts, as rain was scantry and unevenly distributed, the organisms may utilize other water sources.

In polar deserts, occasional snow melts that imbibe the porous rocks seem to be the principal source of water (Friedmann, 1978). In hot deserts nightly dew condensation retained in the rocks may be the main source of water (Friedmann *et al.*, 1967). In case of endoliths, the surface crust of the rocks probably retards evaporation and the porous rock acts as a water trap (Friedmann, 1972). In hypolithic algae, pebbles lying on desert soil retard evaporation with small humidity islands persisting underneath (Vogel, 1955). The physical structure of the microhabitat, the gelatinous sheath of the algae acts in retaining humidity during drought. At the same time desert cyanobacterial species were unable to photosynthesize at lower water potentials and probably react to water stress by cessation of their metabolic activity to rapidly restart again upon rewetting (Potts and Friedmann, 1981). Rock algae were exposed to a wide range of light intensities varying from full sunlight for epilithic algae to very low intensities for endolithic and hypolithic algae. Those algae which were exposed to strong light often have coloured sheath material. Hypo- and endolithic algae live under or in semitranslucent to translucent rocks were they occupy a well defined layer, most probably controlled by the steep gradient of light intensity penetrating into rock or soil. The rocks act as a filter and reduce the light intensity to 30-0.005 per cent of the incidential irradiance in hypolithic algae (Vogel, 1955; Berner and Evenari, 1978) and to 0.1-0.01 per cent, depending on the humidity of the rock, in cryptoendolithic algae (Friedmann and Ocampo-Friedmann, 1984).

Rock algae were eurytherm as their environment undergoes daily and annual temperature variations. In temperate regions, the surface temperature goes up to 50°C in summer while during winter the algae have to survive subzero temperature in contrast to tropical regions where the temperature always remains above the freezing point. In polar regions, algae were frozen for most of the year. During summer, the temperature fluctuates between -10°C and +10°C (McKay and Friedmann, 1985) resulting in daily freeze/thaw cycles. The inorganic nutrients necessary for algal growth were probably found in the soil for hypolithic algae and in the rock for other lithophytic algae. They may also be conveyed to the rocks by rain, snow, and dry atmosphere fallout. The source of nitrogen was of interest. In desert endolithic algal communities biological nitrogen-fixation was non-existent or rare and the source of compound nitrogen was inorganically fixed in the atmosphere, reaching the rock surface by precipitation (rain or snow) or dry fallout. In these communities productivity was low and the incoming fixed-nitrogen results in an accumulation of surplus nitrate in the rocks. Hypolithic algae were known from most deserts and semi-deserts in the world. These were reported from the South-Western United States, South America, Middle East, South Western Africa, Australia, and Antarctica (Vogel, 1955). The algal vegetation was formed by coccoid and filamentous cyanobacteria and green algae which form a conspicuous green zone under the stones.

The endolithic algae survive not by physiological adaptation to low temperatures, but by changing their mode of growth, being able to grow between the crystals of porous rocks. Their activity results in mobilization of iron compounds and in rock withering with a characteristic pattern of expoliation. The inorganic nutrients required by endolithic microorganisms were probably available in the rock substrate (Friedmann, 1982). The endolithic algal community was dominated by a mixed assemblage of coccoid cyanobacteria including *Chroococcidiopsis*, sp., *Gloeothece* sp. with the sarcinoid green algae *Borodinella* and *Fasiculochloris.* The cryptoendolithic algae occurred as uniform well defined bands in light-coloured sandstones and also as scattered patches in dark sandstones. The algal communities varied in generic composition, chlorophyll—a content and location within the different sandstones (Bell *et al.*, 1986). The rocks colonized by cryptoendolithic algae were limestone and sandstone which were primarily light coloured and have a porous structure. Cryptoendolithic algae inhabit the spaces between the particles of porous rocks, at a depth of 1 to several mm where they form a 0.1-2.5 mm wide distinct coloured layer. The algal zone runs at a uniform depth following the contours of the rock surface; the upper and lower boundaries of the zone seem to be determined by light intensity gradient.

Chasmoendolithic algae occupy spaces within rocks that were open to the surface and that range from coarse cracks to microscopic fissures. As the cracks and fissures were in connection with the outer rock surfaces, debris carried by the air can accumulate in the wider spaces and constitute a sort of portsoil which makes the chasmoendolithic habitat similar to the hypolithic (Broady, 1981a). The vegetation forms green belts in the crack running a few mm deep parallel to the surface and was dominated by cyanobacteria.

Euendolithic algae actively penetrate the rock substrate, forming channels of the same contour as the boring organism. Limestone boring algae had been known for a long time in both freshwater and marine habitats where they can play an important role in the destruction of coastal limestone (Golubic *et al.*, 1975). An interesting geological formation, called phytokarst, resulted from boring activities of cyanobacteria (Folk *et al.*, 1973) is its typical example.

Epilithic algae inhabit the exposed rock surfaces as a crust or as a few mm thick layer. They were generally the first photosynthetic organisms that colonize these surfaces (Fritsch, 1907). They occur mostly in habitats where environmental conditions were not too severe. They were infrequent in deserts and seem to be absent in extremely arid habitats. Cyanobacteria followed by green algae were the most important organism of the epitithic vegetation. Light intensity and humidity seem to be the most important factors that determine the composition of an epilithic algal vegetation. Epilithic algae inhabiting surfaces as well as crevices of rocks and stones receive moisture mainly from the atmosphere. Among species present on acidic rocks were genera like *Chroococcus, Gloeocapsa, Stigonema* and *Calothrix*, occurring as blackish masses (Round, 1973). *Pleurococcus vulgaris* (Grev.) Menegh. and *Stichococcus bacillaris* Nag. were the main species, covers the rock surfaces in dry and shadowed places. *Trentepohlia* species often form extensive orange patches. The *Gloeocapsetum* association dominated by species of the cyanobacterial genus *Gloeocapsa* was found in water seepages on steep and bare rock surfaces. This association forms conspicuous black algal crusts that has been characterised as 'Tintenstriche'.

Many microorganisms occurring as epilithophytes excrete organic exopolymers-slimes, which fill up the pores of stones and keep the water fixed to the stone. Water contributes to the weakening of the stone texture. The water draining from the stones decreased the pH value within 6 months and after 24 months a slow increase of the pH value was noted. Microorganisms can only play an important role in deterioration if they were supplied with substrate. The nitric acid which was excreted by the microorganisms as metabolic end-product has been reported to cause severe corrosion (Diercks *et al.*, 1991).

Algae on brick and cement walls that occur on buildings terraces of Allahabad and Varanasi, India were tuft forming species of *Tolypothrix, Scytonema*, intermixed with *Phormidium* and crust forming *Gloeocapsa* generally mixed with *Myxosarcina* while *Porphyrosiphon* was very common on soil surface and bark of mango and other plants.

It was a common observation that many algae occur frequently and spoil the affected external surface of walls, make them ugly, deteriorate the texture of the substratum, by formations of brownish green, blackish papery, greyish brown, black and black powdery patches (Chadha *et al.*, 1980). It was noteworthy that none of these algae form akinetes or spores but continue their existence from season to season. During mid summer months the temperature of terrace goes up to 60-65°C, this coupled with high light intensity and extreme dryness make it most inhospitable to any algae. These subaerial algae survive the severe conditions for about three months and resume their growth after onset of monsoon (Tripathi and Talpasayi, 1980).

Cave algae cover the surfaces of cave walls more and less densely, extends from the entrances of the caves to the end of the photic zone. Geographic location and the configuration of the caves determine the microclimatic conditions of the caves, especially that of temperature and humidity. The microclimate of caves was characterized by even temperatures throughout the year, an even level of humidity, and a pronounced light gradient. The better illuminated wall surfaces near the entrance of caves were generally colonized by green algae and cyanobacteria (Dobat, 1968). Towards the depth of the cave, the algal vegetation becomes less and less dense.. Many limestone cave walls were colonized by calcifying filamentous cyanobacteria which form a greyish, mold-like coat on the substrate. *Scytonema julianum* (Kutz.) Menegh. dominates the vegetation near the cave entrance, whereas *Geitleria calcarea* Friedm. becomes dominant towards the less illuminated areas and forms pure populations at the end of the photic zone. *Chroococcidiopsis kashii* Friedm., a baeocyte forming unicellular cyanobacteria, occurs in dry lime stone caves in Israel (Friedmann, 1961) and seems to tolerate substrates with high nitrate concentrations (Friedmann, 1962). *Speleopogon cavarae* Borzi was another filamentous cyanobacterial species from caves (Borzi, 1917). *Gomontiella magyarina* Claus, a cyanobacterium was reported from caves in Hungary (Claus, 1960) and the United States (Jones, 1964). *Palikiella* (Claus, 1962) and *Baradleia* (Hajdu, 1966), were the two cyanobacterial genera known from caves in Hungary, *Cyanidium chilense* a new species was reported from a cave in Chile (Schwabe, 1936). *Gloeocapsa* NS_4, a cyanobacterium grew in low light levels inside cave entrances shows striking ultrastructure adaptations (Cox *et al.*, 1981). *Loriella osteophila* Borzi was found growing in low light environments, either on fossil coral debris lying on the ground in a dense low land rainforest, or at the entry of a small limestone cave. The greyish calcified filaments were repeatedly dichotomously branched and very fragile (Hoffmann, 1990).

Epiphytic algae commonly occur on higher plants either on tree bark (epipholeic algae) or on leaves (epiphyllic algae). Free living species of epiphytic algae also participate in lichen associations (Tschermak-Woess, 1978). The epiphytic algal vegetation seems to be controlled by physical factors rather than by host plant species. The epiphytic habitat was generally more arid than the soil and was subjected to periodic desiccation. In windy tree bark habitat, atmospheric humidity may be an important source of water supply (Barkman, 1958). Bark fissures may create a special micro-climate as they were shady, retain moisture for a longer period, and were protected against the wind. In epiphloeic algae, nutrients were supplied by the dissolution of salts from the barks. The presence of sporopollenin in some of the epiphytic algae may provide an effective protection against desiccation (Good and Chapman, 1978). The different epiphloeic algae were *Pleurococetum vulgaris* forms a dull green, powdery, water repellant loosely adheres to the bark on the shady side of trees, *Prasioletum crispae,* form a valvety felt on tree barks, especially in alternatively wet and dry rain-tracks and *Trentepohlietum abietinae*, forms orange cushions or crusts on trees, which prefers shady sites (Barkman, 1958).

Desert algae were responsible for the formation of crusts in many areas. Moisture was an important factor in the development of desert crusts. Desert crusts built by *Microcoleus* sp. become thicker, because under favourable conditions, i.e. at water saturation, the filaments were pulsed out of their common sheath and afterwards moved via phtotaxis, leaving behind a lot of sheath material and slime. This material was responsible for the increase in thickness of desert crusts (De Winder *et al.,* 1989). The cyanobacterium *Microcoleus* was a common species in many algal crusts (Brock, 1975). Algal crusts were layers of microorganisms which were subjected to periods of dryness and is a geographically wide spread phenomenon (Campbell, 1979). During drying, they became rigid. These rigid crusts play an important role in erosion process, especially in arid zones. When a crust desiccates its surface become hydrophobic, which would cause a surface run-off during rain. This in turn, leads to an increase in water erosion, on the other hand, wind erosion was reduced. Cyanobacteria were the organisms responsible for the initial formation of aggregates with the sand. The cyanobacteria were surrounded by mucilage, which acts as a cementing agent (Robins *et al.*, 1986). While the crust was being formed, the algae were capable of increasing the total amount of organic matter.

Antarctic algae: Antarctica (77°33' S, 161°00E) can be regarded as the most extreme and inhospitable environment on Earth. This area attracted the interest of microbiologists because the combination of extreme drought and cold poses utmost demands on the adaptive capacity of microorganisms. The exposed ortho-quartzite rocks form the dry valley area of this region revealed the presence of endolithic cyanobacteria *Gloeocapsa* sp. which forms a dark greenish zone approximately 1.5 mm wide parallel to and approximately 1 to 2 mm beneath the rock surface. The rock fabric was porous, and the algae grew in the air spaces between the rock particles. The transparency of the rock allows light to penetrate into the algal zone. The crevices between particles formed a coherent air space system in the rock fabric (Friedmann and Ocampo, 1976).

Ecophysiology of cyanobacteria in extreme environments: Many cyanobacteria possess the ability to withstand extended periods of desiccation and had the capacity to revive from the air-dry state (Brock, 1975; Bewly, 1979 and Whitton *et al.*, 1979). Desiccation tolerance is a widespread and long known phenomenon, occurring over an extremely diverse range of taxa including bacteria, plants and animals (Crowe and Clegg, 1978). The cosmopolitan terrestrial cyanobacterium *Nostoc commune* was also able to tolerate acute were stress and can survive in the air-dry state for many years. Under natural conditions, the cells were embedded and immobilized in a water-absorbing sheath composed of carbohydrates. Growth results in the formation of macroscopic colonies (0.5-3.0 mm thick) which covers areas of several square centimeters. Due to its prokaryotic cellular organization and higher plant type photosynthesis, *Nostoc commune* lends itself as a suitable model system for the study of desiccation tolerance at the molecular level. *Nostoc commune*, a desiccation-tolerant cyanobacterium accumulates a novel group of acidic proteins when colonies were subjected to repeated cycles of drying and rehydration. The proteins occurred in high concentrations, which had a protective function on a structural level (Scherer and Potts, 1989). *Nostoc commune* can withstand prolonged periods of deisccation (years) and recover their metabolic activities rapidly upon cell rehydration. It had been shown in field materials of *Nostoc commune*, as well as in immobilized axenic cultures of *Nostoc commune* UTEX 584, that there was a reproducible sequential recovery of metabolic activities upon rewetting of cells beginning with respiration, then photosynthesis, and finally nitrogen-fixation (Scherer *et al.*, 1986). The first sensitive site affected by desiccation was electron transport between plastoquinone and P 700. On rehydration, recovery was rapid and complete. In contrast, desiccation-intolerant algae did not recover when rehydrated following even limited water loss (Wilkens *et al.*, 1978) suggesting irreversible chloroplast damage. Desiccation-tolerant terrestrial algae gradually cease to respire and photosynthesize during drying, recovery was on subsequent rehydration. It has been suggested that the accumulation of fats or oils in the protoplasm of some algae (may be during water loss) increases their capacity to withstand desiccation (Collyer and Fogg, 1955). Algae produce pigments (e.g. carotenoids) for photoprotection, i.e. to act as filters to reduce the levels of irradiation impinging on sensitive structures within the cell. Some algae survive slow desiccation better than rapid desiccation (Evans, 1959). Therefore, the observation that some desiccation-tolerant species possess thick walls and can produce mucilage may have some significance, both could retard water loss from cells in a drying environment (Bewley, 1979). The mechanisms of desiccation tolerance in *Nostoc commune* was found out by demonstrating that both phospholipid bilayers and proteins can be stabilized during water stress by sugars, especially by trehalose (Crowe *et al.*, 1987). Trehalose was found in a variety of microoganisms including cyanobacteria when they were subjected to drying (matric water stress) or osmotic stress.

Nostoc commune shows a marked capacity for desiccation tolerance. Reserve materials were maintained intact during desiccation and as such, they may provide the substrates required for the recovery of cell function upon rewetting. As the heterocysts were evidently damaged by desiccation, the time taken for the recovery of nitrogen-fixation may reflect the time needed for new heterocysts to be differentiated (Peat and Potts, 1987). *Chroococcus* N 41, a cryptoendolith isolated from a hot desert rock and *Chroococcus* S 24, a marine from were of coccoid cyanobacteria that can survive prolonged desiccation and can revive when rewetted (Potts *et al.*, 1983). The sequentional colonization of organisms is likely to be determined by the ability of organisms to respond to the demanding environmental conditions. The response towards rewetting was found to greatly depend on the matrix potential of the supporting substratum and on the organism proper (De Winder *et al.*, 1989). Cyanobacteria were also known to form crusts in desert or tropical soils (Roger and Reynaud, 1982). In such environments they thrive under conditions of low matrix water potential. *Cirnalium epipsammum* is such a new species of filamentous cyanobacterium whose trichomes are composed of elliptical cells. The organism is non-motile and drought resistant, and its cell surface is hydrophillic. The cell is relatively thick and contains poly-B (1, 4) glucan (cellulose), which is unusual for cyanobacteria (De Winder *et al.*, 1990).

In several cyanobacteria, glycogen, had been identified as granules stored between thylakoids (Allen, 1984). Other carbohydrates were found in the sheath, which forms a fibrillar and mucilaginous structure layered on to the outside of the gram-negative cell wall (Drews and Weckesser, 1982; Abdelahad and Bazzichelli, 1984 and Adhikary *et al.*, 1986). Mucilage acts as means of protection against stress (water, heat). *Nostoc commune*, a drought tolerant terrestrial cyanobacterium forming leaf-like mats on rocks and soil, which survives as vegetative cells, showing net gain in its water buget and exhibiting nitrogen-fixation during short rewetting periods (Coxson and Kershaw, 1983 b). When hydrated, the sheath contributes substantially to the increase in weight and volume of the colonies (Scherer *et al.*, 1984). A capacity to tolerate water stress and desiccation was a feature of many cyanobacteria of terrestrial origin (Potts and Friedmann, 1981). Studies concerned with the water relations of cyanobacteria dealt primarily with the response of these microorganisms to elevated solute concentrations (Walsby, 1982). Resistance towards extreme water stress, i.e. desiccation was widely observed in plant kingdom, preferably in lichens and algae. The response of the terrestrial cyanobacteria *Nostoc flagelliforme, Nostoc commune* and *Nostoc* sp. to water uptake had been investigated after a drought period of 2 years. Rapid half-times of rewetting (0.6, 3.3, and 15.5 min respectively) were found (Scherer *et al.*, 1984). The surface to mass ratio of the three species was inversely correlated to the speed of water uptake and loss. Respiration started immediately after a 30 min rewetting period, whereas photosynthetic oxygen evolution reached its maximum activity after 6 and 8 h with *Nostoc commune* and *N. flagelliforme* respectively. In dark, recovery of oxygen uptake by *N. commune* was somewhat impaired, while slightly stimulated with *N. flagelliforme.* With both species, recovery of photosynthesis was inhibited in darkness (Scherer *et al.*, 1984). Thus, after a 2 year drought period, the physiological sequence of reactivation was respiration—photosynthesis-nitrogen-fixation. Respiration and photosynthesis preceded growth and were exhibited by existing vegetative cells, whereas recovery of nitrogen-fixation was dependent on newly differentiated heterocysts (Scherer *et al.*, 1984). Terrestrial organisms facing drought have, an addition to osmotic forces, also withstand matric water potential stress, which includes both adsorption and capillary effects in the solid phase of their physical environment. This matric potential constitutes the major component of the water potential in most water unsaturated soils. Organisms exposed to changing matric water potential not only have to overcome periods of complete drought, but also cope with the demanding process of desiccation and rehydration. The possible mechanisms aiding to withstand osmotic stress is that drought tolerant organisms were often embedded in an environment such as fine textured sand stone and/or possess a thick mucilaginous sheath around their trichome/cells (Campbell, 1979).

Cyanobacteria, those known to inhabit terrestrial environments are often subjected to intense solar radiation. They were often the primary colonizers of bare rock surfaces and soils (Whitton, 1987). Under such conditions, they must possess efficient mechanisms to prevent or to counteract the harmful effects of strong solar radiation, including those caused by the particularly deleterious UV region of the spectrum. Ultraviolet (< 400 nm) was an integral part of solar radiation. Based on wavelength, ultraviolet radiation was subdivided into three regions, namely UV-A (315-400 nm), UV-B (280-315 nm) and UV-C (> 280 nm). Amongst these, the UV-B radiation was severely affected by reduction in stratospheric ozone. The influence of solar UV on biological and other phenomena depends not only on ozone layer thickness and sun's angle, but also on the nature of the biological effects themselves. The damaging effects of enhanced UV-B radiation resulted from a possible destruction of the stratospheric ozone layer by man-made chemicals, such as chloroflurocarbons (Biggs *et al.*, 1981; Gold and Cladwell, 1983; Flint and Cladwell, 1984 a and Renger *et al.*, 1989). UV-B induced reductions of leaf area, fresh and dry weight, lipid content and photosynthetic activity in higher plants and its effect on many metabolic processes, pigmentation, and community composition of biological systems other than higher plants have been reported. Cyanobacteria, is a group that occupied an important place in both aquatic and terrestrial habitats. This was because these organisms constitute phylogenetically the oldest group of oxygen-evolving photosynthetic prokaryotes and it seems that during their evolutionary history, they had faced more intense solar UV radiation than other organisms. Cyanobacteria grow in diverse habitats ranging from hot springs to the arctic and therefore they were expected to face differential levels of UV-B. In addition to these properties, some of these organisms fix atmospheric nitrogen and enrich the fertility of paddy fields and other soils consequently, any impact of UV-B radiation on diazotrophic cyanobacteria would be expected to affect the productivity in such habitats. Diazotrophic enzyme in cyanobacteria was more sensitive to UV-B than other metabolic processes (Newton *et al.*, 1979). UV-B radiation was generally inhibitory for several physiological processes and particularly to the photosynthetic components (Tyagi *et al.*, 1991). Mechanisms that counteract UV-mediated photodamage in cyanobacteria include both dark DNA repair and photorepair (Levine and Thiel, 1987), the presence of detoxifying enzymes (Mittler and Telor, 1991) and the action of photoprotective carotenoids (Buckley and Houghton, 1976). Among these possible mechanisms to prevent UV-photodamage were negative photomovements or avoidance responses and the synthesis of UV sunscreen compounds (Garcia-Pichel and Castenholz, 1991). In addition, the presence of UV-absorbing compounds in cyanobacteria, that had also been playing a UV-photo protective role, is the extracellular sheath pigment, scytonemin.

Scytonemin, the yellow-brown pigment of cyanobacterial extracellular sheaths, was found in species exposed to intense solar radiation. Scytonemin occurred predominantly in sheaths of the outermost parts or top layers of cyanobacterial mats, crusts, or colonies. The sheath pigmentation was first noted by Nageli, who coined the term 'scytonemin' in 1877. It was lipid soluble and had a prominent absorption maximum in the near ultra-violet region of the spectrum with a long tail extending to the infrared region. Scytonemin absorbs most strongly in the UV-A spectral region (315-400 nm with peak at 384 nm) with an in vivo λ max = 370 nm. There was also significant absorbance in the violet and blue region as well as in the UV-B (280-320 nm) and UV-C (190-280 nm) range. The pigment had been isolated and identified from more than 30 species of sheathed cyanobacteria from diverse geographic regions, including fresh water, terrestrial, and marine habitats, whenever exposure to strong solar irradiance occurs (Garcia-Pichel *et al.*, 1992). The absorption properties of scytonemin was a proof of its effectiveness as a UV screen. It has been shown that common levels of scytonemin in cyanobacterial sheaths may be sufficient to prevent 85-90 per cent of incident UV-A radiation from entering the cells.

Scytonemin, an effective, photo-stable, dimeric molecule (mol. wt. 544) of indolic and phenolic sub-units and is known only from the sheaths enclosing the cells of cyanobacteria. Scytonemin is formed from a condensation of tryptophan and phenylpropanoid derived subunits. The linkage between these units is unique among natural products and this novel ring structure is formed the "scytoneman skeleton". This pigment had been shown to provide significant protection to cyanobacteria against damage by ultraviolet radiation.

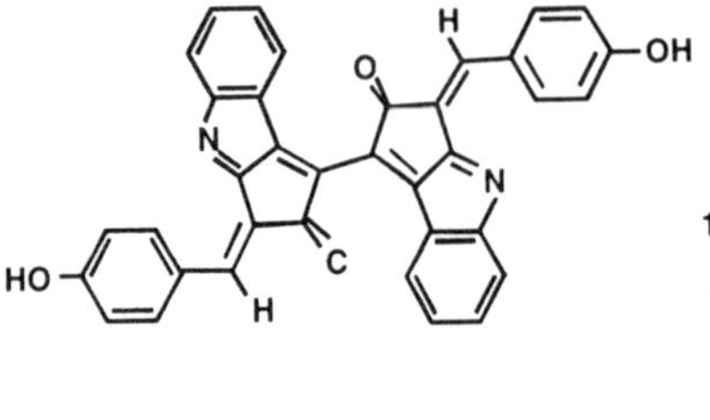

Scytonemin 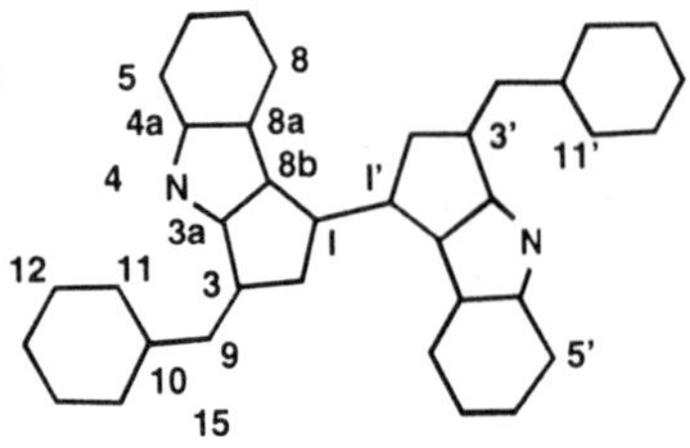 **Scytoneman Skeleton**

The pigment occurred in all sheathed cyanobacteria and represents UV screening strategy far more ancient than that of plant flavonoids and animal melanins (Proteau *et al.*, 1993).

A variety of colourless, water soluble compounds known as mycosporine like amino acids (MAAS) that had been characterized in fungi, eukaryotic algae, corals, and starfish, had been found in a wide number of cyanobacteria. Chemically, MAAS contain a substituted cyclohexenone linked with an amino acid (or its imino alcohol). They present a characteristic single absorption band in the UV spectrum, between 230-400 nm, with maxima ranging from 310-360 nm, and absorb only weakly below and above the main band (Garcia-Pichel and Castenholz, 1993). UV radiation can trigger the production and accumulation of mycosporines in fungi (Trione *et al.*, 1969) corals, (Jokiel *et al.*, 1982) a dinoflagellate, (Carreto *et al.*, 1990) and cyanobacteria (Garcia-Pichel and Castenholz, 1990). Ultraviolet radiatioon causes damage to photosystem II. The decrease in photosynthetic activity that occurs when a plant was exposed to

UV or high irradiances of visible light is termed photoinhibition (Tyystjarvi, 1993). Short term photo-inhibition by UV-radiation can reduce primary production by phytoplankton. Phytoplankton showed great variation in sensitivity to UV radiation due to the presence of different mechanisms of their escape or protection from damaging effect of UV radiations.

The increase in ultraviolet-B (UV-B: O, 0.290-0.320 μm) radiation received by plants due to stratospheric ozone depletion hightens the importance of understanding UV-B tolerance. Photosynthetic tissue is to be protected from UV-B radiation by UV-B absorbing pigments/compounds (e.g. flavonoids) (Middleton and Teramura, 1993), which were demonstrated in several species of terrestrial algae.

ECOLOGICAL SIGNIFICANCE OF TERRESTRIAL ALGAE AND CYANOBACTERIA

The present day success and presumably also in the earlier geological periods is no doubt of a number of features widespread in these lower group of photosynthetic organisms. The temperature optimum for many or most algae and cyanobacteria is higher by atleast several degrees than most of eukaryotic plants. Many terrestrial species of algae tolerate high levels of ultraviolet irradiation and also tolerate desiccation and water stress. The ability of many species of cyanobacteria to fix atmospheric nitrogen provides a competitive advantage where levels of combined nitrogen is low. Many terrestrial algae also form characteristic crust on soil and sand helping to check soil erosion. However, few are aesthetic nuisance when occur as crust on the exposed surfaces of temples, monuments and art works. In addition their growth is often associated with secretion of substances that leads to withering of the monuments of archieological importance. Understanding the biodiversity of terrestrial cyanobacteria and algae forms in several ecological niches is important to develop management models for (i) checking soil/sand erosion (ii) use of sunscreen pigments of cyanobacteria on UV protective lotions/ointments for preventing skin injury due to seamingly increase of UV-B radiation in the globe on the avent of stratospheric ozone depletion (iii) characterization of extra cellular substances leached by these organism which has a direct bearing on the life of the monuments and to develop control measures for protection of these gigantic architectures from destruction,

(iv) understanding nitrogen fixing ability of cyanobacterial crusts in the upland fields and their contribution to fixed nitrogen for sustainability of crop production.

References

Abdelahad, N. and Bazzichelli, G., 1984. An electron microscopic investigation on granules observed in the sheath of *Nostoc commune* Vauch. Arch. Hydrobiol. Suppl. 67: 205-211.

Adhikary, S.P., 1986. Production of giant form in *Westiellopsis prolifica* under heterotrophic culture conditions. Curr. Sci. 55: 669-670.

Allen, M.M., 1984. Cyanobacterial cell inclusions. Ann. Rev. Microbiol. 38: 1-25.

Barkman, J.J., 1958. Phytosociology and ecology of cryptogamic epiphytes. Koninklijke van Gorcum & comp. N.V.G.A. Hak and Dr. H.J. Prakke, Assen, The Netherlands, pp. 628.

Bastia, A.K., Satapathy; D.P. and Adhikary, S.P., 1993 Heterotrophic growth of several filamentous blue-green algae. Archiv. Hydrobiol. 70: 65-70.

Bell, R.A., Athey, P.V. and Sommerfeld, M.R., 1986. Cryptoendolithic algal communities of the Colorado Plateau. J. Phycol. 22: 429-435.

Berner, T. and Evenari, M., 1978. The influence of temperature and light penetration on the abundance of the hypolithic algae in the Nagev Desert of Israel. Oekologia 33: 255-260.

Bewley, J.D., 1979. Physiological aspects of desiccation tolerance. Ann. Rev. Plant Physiol. 30: 195-238.

Biggs, R.H., Kossuth, S.U. and Teramura, A.H. 1981. Response of 19 cultivars of soyabeans to ultraviolet-B irradiance. Physiol. Plant. 53: 19-26.

Booth, W.E., 1941. Alge as pioneers in plant succession and their importance in erosion control. Ecology 33: 38-46.

Borzi, A., 1917. Studi sulle Mixofiecee. Nuovo Giorn. Bot. Ital. 24: 100-112.

Broady, P.A. 1981a. The ecology of chasmolithic algae at coastal locations of Antarctica. Phycologia 20: 259-272.

Brock, T.D., 1975. Effect of water potential on a *Microcoleus* (cyanophyceae) from a desert crust. J. Phycol. 11: 316-320.

Buckley, C.E. and Houghton, J.A., 1976. A study of the effects of near UV-radiation on the pigmentation of the blue-green alga *Gloeocapsa alipicola*. Arch. Microbiol. 107: 93-97.

Campbell, S.E., 1979. Soil stabilization by a prokaryotic desert crust: implications for precambrian land biota. Origins of Life. 9: 335-348.

Carreto, J.I., Carignan, M.O., Daleo, G. and De Marlo, S.C., 1990. Occurrence of mycosporine-like amino acids in the red tide dinoflagellate *Alexandrium excavatum*: UV-protective compounds. J. Plankton Res. 12: 909-921.

Chadha, A., Pandey, D.C. and Tiwari, G.L., 1980. Investigations on the algae inhabiting the walls of buildings, part-I General Survey. In: Proceedings National Workshop on algal systems. Eds. Seshadri, C.V., Thomas, S. and Jeejibai, N. Indian Institute of Technology, Madras, pp. 238-240.

Claus, G., 1960. Re-evaluation of the genus *Gomontiella*. Rev. Algol. N.S., 5: 103-111.

Claus, G., 1962a. Data on the ecology of the àlgae of Peace Cave in Hungary. Nova Hedwigia, 5: 55-80, Pl. 34-36.

Collyer, D.M. and Fogg, G.E., 1955. Studies on fat accumulation by algae. J. Exp. Bot. 17: 256-275.

Cox, G. Benson, D., Dwarte, D.M., 1981. Ultra structure of a cave wall Cyanophyte-*Gloeocapsa* NS4. Arch. Microbiol. 130: 165-174.

Coxson, D.S. an Kershaw, K.A., 1983b. Rehydration response of nitrogenase activity and carbon fixation in terrestrial *Nostoc commune* from Stipa-Bouteloa grassland. Can. J. Bot. 61: 2658-2668.

Crowe, J.H. and Clegg, J.S., 1978. Dry Biological Systems. Academic Press, New York.

Crowe, J.H., Crowe, L.M., Carpenter, J.F. and Wistrom, C.A., 1987. Biochem. J. 242: 1-10.

De Winder, B.D. Matthijs, H.C.P. an Mur, L.R., 1989. The role of water retaining substrate on the photosynthetic response of three drought tolerant phototrophic microorganisms isolated from a terrestrial habitat. Arch. Microbiol. 152: 458-462.

De Winder, B.D., Pluis, J., Reus, L.D. and Mur, L.R. 1989. Characterization of a cyanobacterial Algal crust in the coastal dunes, of the Netherlands, Amer. Soc. Microbiol. 77: 83.

De Winder, B.D., Matthijs, H.C.P. and Murr, L.R. 1990. The effect of dehydration and ion stress on carbon dioxide fixation in drought-tolerant phototrophic microorganisms, FEMS Microbiol. Ecology 74: 33-38.

De Winder, B.D. Stal, L.J. and Mur, L.R. 1990. *Crinalium epipsammum* sp. nov: a filamentous cyanobacterium with trichomes composed of elliptical cells and containing poly. β- (1, 4) glucan (cellulose). J.G. Microbiol. 136: 1645-1653.

Diercks, M., Sand, W. and Bock, E., 1991. Microbial corrosion of concrete. Experentia 47: 514-516.

Dobat, K., 1968. Die Pflanzen-und Tierwett der charlottenhohle. Karsu-u. Hohlenkunde, Reihe A, 3: 37-50.

Drews, G. and Weckesser J., 1982. Function, structure and composition of cell walls and external layers. In: The biology of cyanobacteria. Eds. Carr, N.G. and Whitton, B.A. Blackwell Scientific Publications, Oxford London Edinburg Boston Melbourne, pp. 333-357.

Durrell, L.W. and Shields, L.M., 1961. Characteristics of soil algae relating to crust formation. Trnas. Amer. Microscop. Soc. 80: 73-79.

Evans, J.H., 1959. The survival of fresh water algae during dry periods. II. Drying experiments. J. Ecol. 47: 55-81.

Flint, S.D. and Cladwell, M.M., 1984a. Comparative Sensitivity of binucleate and trinucleate pollen to ultraviolet radiation: A theoretical perspective. In: Stratopheric Ozone Reduction, Solar Ultraviolet Radiation and Plant Life, Eds.Worrest, R.C. and Cladwell, M.M., Springer-Verlag, Berlin.

Folk, R.L., Roberts, H.H., and Moore, C.H., 1973. Black phytokarst from Hell, Cayman Islands, Britain West Indies. Bull. Geol. Soc. Amer. 84:2351-2360.

Friedmann, I., 1961. *Chroococcidiopsis kashaii* sp. n. and the genus Chroococcidiopsis

(Studies on cave algae from Israel. III.) Oesteer, Bot. Z. 108: 354-367.

Friedmann, E.I., 1962. The ecology of the atmophytic nitrate-alga *Chroococcidiopsis Kashaii* Friedmann. (Studies on cave algae from Israel IV). Arch. Mikrobiol. 42: 42-45.

Friedmann, E.I., 1972. Ecology of lithophytic algal habitats in Middle Eastern and North American desert. In: Ecophysiological foundation of ecosystem productivity in arid zones. Ed. Rodin, L.E. Nauka; USSR. Acad. Sci., Leningrad, pp. 182-85.

Friedmann, E.I., 1978 Melting snow in the dry valleys is a source of water for endolithic microorganisms. Antarct. J.U.S. 13: 162-63.

Friedmann, E.I., 1982, Endolithic microorganisms in the Antarctic cold desert. Science 215: 1045-53.

Friedmann, E.I., and Galun, M. 1974. Desert algae, lichens, and fungi. In: Desert biology 2. Ed. Brown, G.W., Academic Press, Inc., New York, pp. 165-212.

Friedmann, E.I., and Ocampo, R., 1976. Endolithic blue-green algae in the dry valleys: Primary producers in the Antarctic Desert ecosystem. Science 193: 1247-1249.

Friedmann, E.I. and Ocampo-Friedmann, R., 1984. Endolithic microorganisms in extreme dry environments: Analysis of a lithobiontic microbial habitat: In: Current perspectives in microbial ecology. Eds. Klug, M.J. and Reddy, C.A. American Soc. Microbiol., Washington, D.C. pp. 177-185.

Friedmann, E.I., Lipkin, Y. and Paus, R.O., 1967. Desert algae of the Nagev (Israel). Phycologia 6: 185-200.

Friedmann, E.I., M.C. Kay, C.P. and Nienow, J.A., 1987. The cryptoendolithic microbial environment in the Ross Desert of Antarctica: Satellite transmitted continuous nanoclimate data, 1984 to 1986. Polar Biol. 7: 273-287.

Fritsch, F.E. 1907. A general consideration of the subaerial and freshwater algae of Ceylon. In: Proc. Roy. Soc. London 79: 197-254.

Garcia-Pichel, F. and Castenholz, R.W., 1991. Characterization and biological implications of scytonemin a cyanobacterial sheath J. Phycol. 23: 395-409.

Garcia-Pichel, F. and Castenholz, R.W., 1993. Occurrence of UV absorbing, mycosporine like compounds among cyanobacterial isolates and an estimate of their screening capacity. Appl. Environ. Microbiol. 59: 163-169.

Garcia-Pichel, F., Shery, N.D. and Castehholz, R.W. 1992. Evidence of for an ultraviolet sunscreen roe of the extracellular pigment scytonemin in the terrestrial cyanobacterium *Chlorogloeopsis* sp. Photochem. Photobiol. 56: 17-23.

Golubic, S. Perkins, R.D. and Lukas, K.J., 1975. Boring microorganisms and microborings in carbonate substrates. In: The study of trace fossils. Ed. Fery, R.W. Springer-Verlag, New York, pp. 229-259.

Good, B.H. and Chapman, R.L., 1978. The ultrastructure of *Phycopeltis* (Chroolepidaceae: Chlorophyta). I. Spronopollenin in the cell walls. Amer. J. Bot. 65: 27-33.

Hajdu, L. 1966. Algological Studies in the caves of Matyas Mount, Budapest, Hungary, Int. J. Speleol. 2: 137-149.

Hoffmann, L. 1990. Rediscovery of *Loriella osteophila* (cyanophyceae) Br. Phycol. J. 25: 391-395.

Jokiel, P.L., and York, R.H. 1982. Solar ultraviolet photobiology of the reef coral

Pocillopora damicornis and symbiotic zooxanthellae. Bull. Mar. Sci. 32: 301-315.

Jones, H.J., 1964. Algological investigations in Mammoth Cave, Kentucky, Int. Speleol. 1: 491-516.

Khoja, T. and Whitton, B.A., 1975. Heterotrophic growth of filamentous blue-green algae. Br. Phycol. J. 10: 139-148.

Levine, E. and Thiel, T., 1987. UV-inducible DNA repair in the cyanobacteria *Anabaena* spp. J. Bacteriol. 169: 3988-3993.

Lipman, C.B., 1941. The successful revival of *Nostoc commune* from a herbarium specimen eighty-seven years old. Bull. Torrey Bot. Club. 68: 664-666.

McKay, and Friedmann, E.I., 1985. The cryptoendolithic microbial environment in the Antarctic cold desert. Temperature variations in nature. Polar Biol. 4: 19-25.

Metting, B., 1981. The systematics and ecology of soil algae. Bot. Rev. 47: 195-312.

Middleton, E.M. and Teramura, A.H., 1994. Potential problems in the use of cellulose diacetate and mylar filters in UV-B radiation studies. Photochem. Photobiol. 57: 744-751.

Mittler, R. and Tel-Or, E., 1991. Oxidative stress responses in the unicellular cyanobacterium *Synechococcus* PCC 7942 Free. Rad. Res. Comm. 12-13; 845-850.

Newton, J.W., Tyler, D.D. and Slodki, M.E., 1979. Effect of ultraviolet-B (280-320 nm) radiation on blue-green algae (cyanobacteria): Possible biological indicators of stratospheric ozone depletion. Appl. Environ. Microbiol. 37: 1137-1141.

Peat, A. and Potts, M., 1987. The ultrastructure of immobilised desiccated cells of the cyanobacterium *Nostoc commune* UTEX 584. FEMS Microbiol. Lett. 43: 223-227.

Peterson, J.B., 1935. Studies on the biology and taxonomy of soil algae. Dansk. Bot. Ark. 8: 1-180.

Potts, M. and Friedmann, E.I., 1981. Effects of water stresss on cryptoendolilthic cyanobacteria from Hot desert rocks. Arch. Microbiol. 130: 257-2171.

Potts, M., Ocampo-Friedmann, R., Bowman, M.A. and Tozun, B., 1983. *Chroococcus* S24 and *Chroococcus* N41 (cyanobacteria): Morphological, biochemical and genetic characterization and effects of water stress on ultrastructure. Arch. Microbiol. 135: 81-90.

Proteau, P.J., Gerwick, W.H., Garcia-Pichel, F. and Castenholz, R. 1993. The structure of Scytonemin, an ultraviolet sunscreen pigment from the sheaths of cyanobacteria. Experientia 49: 825-829.

Renger, G., Volker, M., Eckert, H.J., Formme, R. Hohn-viet, S. and Graber, P., 1989. On the mechanisms of photosystem II deterioration by UV-B irradiation. Photochem. Photobiol. 49 97-105.

Robins, R.J., Hall, D.O., Shi, D.J. Turner, R.J. and Rhodes, M.J.C., 1986. Mucilage acts to adhere cyanobacteria and cultured plant cells to biological and inert surfaces. FEMS Microbiol. Lett. 34: 155-160.

Roger, P.A. and Reynaud, P.A., 1982. Free-living blue-green algae in tropical soils. In: Microbiology of tropical soil and plant productivity. Eds. Dommergues, Y.R. and Diem, H.G. The Hauge: Martinus Nijhoffl Dr. W. Junk Publishers, pp. 147-168.

Round, F.E., 1973. The biology of the algae. 2nd ed. London: Arnold.

Scherer, S. and Potts, M. 1989. Novel water stress protein from a desiccation-tolerant cyanobacterium. J. Biol. Chem. 264: 12546-12553.

Scherer, S., Ernst, A., Chen, T.W. and Boger, P., 1984. Rewetting of drought-resistant blue-green algae: Time course of wtaer uptake and reappearance of respiration, photosynthesis, and nitrogen-fixation Oecologia 62: 418-423.

Scherer, S., Chen, T.N. and Boger, P. 1986. Recovery of adenine-nucleotide pools in terrestrial blue-green algae after prolonged drought periods. Oecologia (Berlin) 68: 585-588.

Schwabe, G.H. 1936. Uber einige Blaualgen aus dem mitteren und sudlichen chile. verh. Deutsch. Wissenschaftl. ver. Santiago (Chile), N.F. 113-174.

Shields, L.M. and Durrell, L.W. 1964. Algae in relation to soil fertility. Bot. Rev. 30: 92-128.

Stainer, R.Y., 1973. Autotrophy and heterotrophy in unicellular blue-green algae. In: The biology of blue-green algae. Ed. Carr, N.G. and Whitton, B.A. Bot. Monogr, 9. Blackwell Scientific Publications, Oxford, pp. 501-518.

Tripathi, S.N. and Talpasayi, E.R.S. 1980. Sulphydryls and survival of subaerial ble-green algae. Curr. Science 49: 131-132.

Trione, J.E. and Leach, C.M., 1969. Light-induced sporulation and sporogenic substances in fungi. Phytopathol. 59: 1077-1083.

Tschemark-Woess, E., 1978. *Myrmecia reticulata* as a phycobiont and free-living-Free-living *Trebouxia*. The problem of *Stenocybe septata*. Lichenologist. 10: 69-79.

Tyagi, R., Kumar, H.D., Vyas, . and Kumar, A. 1991. Effects of Ultraviolet-B radiation on growth, pigmentation NaH^{14} Co_3 uptake and nitrogen metabolism in *Nostoc musccorum*. In: Impact of global climatic changes on photosynthesis and productivity, New Delhi, India, ICAR and Far Eastern Regional Reserach Office of the United States Dept. of Agriculture, pp. 109-124.

Tyystjarvi, E., 1993. Photoinhibition-struggle between damage and repair of photosystem II. Thesis. Grafia Oy. Turku. ISBN 952-90-4519-0.

Vogels, S., 1955. Niedere "Fensterpflazen" in der sudarfrikanischen Wuste. Beitr. Biol. Pflanzen 31: 45-135.

Walsby, A.E., 1982. Cell-water and cell-solute relations. In: The biology of cyanobacteria. Eds. Carr, N.G. and Whitton, B.A. Blackwell Scientific Publications, Oxford, pp. 237-262.

Whitton, B.A., 1987. Survival and dormancy of blue-green algae. In: Survival and dormancy of microorganisms. Eds. Heins, Y. John Wiley and Sons, New York, pp. 109-167.

Whitton, B.A. Donaldson, A. and Potts, M., 1979. Nitrogen-fixation by *Nostoc* colonies on terrestrial environments of Aldabra Atoll, Indian Ocean. Phycologia 18: 278-287.

Wilkens, J., Schruber, U. and Vidaver, W., 1978. Chlorophyll fluorescence induction: An indicator of photosynthetic activity in marine algae undergoing desiccation. Can. J. Bot. 56: 2787-2794.

8

Controlling Water Pollution

DILIP BISWAS

India is enviably endowed in respect of water resources. The country is literally criss-crossed with rivers and blessed with high precipitation mainly due to the south-west monsoon which accounts for 75% of the annual rainfall. The major river basins along with the medium and minor river basins account for almost 91% of the country's entire drainage area. Yet, inspite of the nature's bounty, paucity of potable water is an issue of national concern.

Potable water has all along been so readily available that few people consider its abundance a miracle. The decline in quantity and deterioration of water quality are directly attributable to unabated competing demands for various sectors of water users and disposal of wastes from different sources including human settlements, industry and agricultural activities.

Like several other developing countries, India is in the midst of massive urbanisation. The proportion of urbanisation in the first forty years of the century was less than 12% and it has risen to 25.72% in the 1991 census. Besides urbanisation, the 1991 census also indicated rapid growth of industrial towns as a conspicuous feature. With the runaway increase in population and urbanisation along with associated activities, water is becoming a scarce resource.

The symptoms of strain in terms of depletion and pollution of water due to unregulated multiple uses of river stretches are already evident. Hence, it is important that a rationale is adopted for maintaining sustainable balance between the quality and quantity of water use.

TABLE 1

Primary water quality criteria for various uses* of fresh waters, as laid down by the Central Pollution Control Board

	Characteristics	*A**	*B**	*C**	*D**	*E**
1.	Dissolved oxygen (DO), mg/1, Min.	6	5	4	4	—
2.	Biochemical oxygen demand (BOD). mg/1, max.	2	3	3	—	—
3.	Total coliform organism**, MPN/100 ml, Max.	50	500	5,000	—	—
4.	pH value	6.5-8.5	6.5-8.5	6-9	6.5-8.5	6.5-8.5
5.	Free ammonia (as N), mg/1, Max.	—	—	—	1.2	—
6.	Electrical conductivity, micromhos/cm, Max.	—	—	—	—	2,250
7.	Sodium absorption ratio, Max.	—	—	—	—	26
8.	Boron, mg/1, Max.	—	—	—	—	2

* Use class: (A) drinking water source without conventional treatment but after disinfection, (B) outdoor bathing organized, (C) drinking water source with conventional treatment followed by disinfection, (D) propagation of wildlife, fishries, (E) irrigation, industrial cooling, controlled waste disposal. **If the coliform is found to be more than the prescribed tolerance limits, the criteria for coliforms shall be satisfied if not more than 20 percent of samples show more than the tolerance limits specified, and not more than 5 per cent of samples show value more than 4 times the tolerance limits. There should be no visible discharge of domestic and industrial wastes into Class A waters. In case of Class B and C the discharge shall be so regulated/treated as to ensure maintenance of the stream standards.

WATER MANAGEMENT

There are several agencies at the Centre and at the State level to deal with surface water, ground water, drinking water supply and sewage disposal. The Central agencies which are concerned with different aspects of water management include the following:

TABLE 2

Details of Major Rivers of India

Sl. No.	*Name of the River*	*Length in India (cm)*	*Basin Area in India (sq. km.)*	*Average annual discharge (MCM)*	*Place of origin*	*Destination*
1.	Ganga	2525	861404	493400	Gangotri glacier, Uttar Kashi, U.P.	Bay of Bengal
2.	Indus	1270	321290	41955	Near Mansarover Lake, Tibet	Arabian Sea
3.	Godavari	1465	312812	105000	Nasik, Maharashtra,	Bay of Bengal
4.	Krishna	1400	258948	67675	Mahabaleshwar, Maharashtra	Bay of Bengal
5.	Brahmaputra	720	187110	510450	Kailash Range, China	Bay of Bengal
6.	Mahanadi	857	141600	66640	Raipur, M.P.	Bay of Bengal
7.	Narmada	1312	98796	40705	Amarkantak, M.P.	Arabian Sea
8.	Cauvery	800	87900	20950	Coorg, Karnataka	Bay of Bengal
9.	Tapi	724	65145	17982	Batul, M.P.	Gulf of Khambat
10.	Pennar	597	55213	3238	Chennakesva Hills, Karnataka	Bay of Bengal
11.	Brahmani	800	39033	18310	Ranchi, Bihar	Bay of Bengal
12.	Mahi	533	34842	8500	Ratlam, M.P.	Gulf of Cambey
13.	Sabarmati	300	21674	3200	Aravli Hills, Gujarat	Gulf of Cambey

1. Ministry of Water Resources :
   - — Central Water Commission,
   - — Central Ground Water Board,
   - — Central Board for Irrigation and Power.
2. Ministry of Urban Development :
   - — Central Public Health and Environmental Engineering Organisation.
3. Ministry of Rural Development.
4. Ministry of Agriculture:
   - — Indian Council of Agricultural Research.
5. Ministry of Environment & Forests :
   - — Central Pollution Control Board,

— Ganga Project Directorate/National River Conservation Directorate,
— Environmental Impact Assessment Wing.

ROLE OF CPCB

As an essential pre-requisite for water quality management, the Central Pollution Control Board has demarcated various stretches of the major rivers (and their tributaries) based on their designated best use, and laid down the water quality criteria using selected primary water quality parameters (Table 1). Alongside, over the years, the Central Board has been involved in the following activities:

* Establishment of a network comprising 480 stations so far set up to monitor water quality of mainly fresh surface water bodies (rivers, lakes) under the Monitoring of Indian National Aquatic Resources (MINARS) and the Global Environmental Monitoring System (GEMS) under WHO programme.
* Launching a ground water quality network of 134 stations to monitor ground water quality in 21 problem areas of the country.
* Mapping of coastal stretches and coastal water quality monitoring in 173 locations.
* River basin-wise surveys (Table 2) for assessment of existing status, pollution potential (Table 3) and action points. The data generated from such surveys provided the inputs for formulation of the Ganga Action Plan and the National River Conservation Plan.
* Assistance to the Ganga Project Directorate in identification of polluted stretches and for conducting water quality monitoring to ascertain the impacts of programmes under the Action Plans.
* In-depth studies of pollution potential in different types of industries and formulation of Minimal National Standards (MINAS) for enforcement through the State Pollution Control Boards and Pollution Control Committees in Union Territories.
* Technical assistance to the Ministry of Environment and

Forests in environmental impact assessment of projects with particular reference to impact on water quality.

CHALLENGES

* Poor sanitation conditions.
* Unchecked disposal of municipal sewage into surface water.
* Paucity of fund and infrastructural facilities in local bodies.
* Non-availability of minimum river flow to utilise assimilative capacity.
* Faulty land-use.
* Operation and maintenance problem in end-of-the-pipe technology for treating industrial waste.
* Obsolete technology and scale of operation in Small Scale Industrial (SSI) units.
* Delay in judicial proceedings for prosecution of defaulting units.

STRATEGY

* Control of pollution at sources giving due regard to techno-economic feasibility.
* Optimal utilisation of assimilative capacity of recipient water bodies to minimise investment on pollution control.
* Maximization of reuse/recyle of waste-water for agricultural/industrial purposes.
* Minimization of pollution control requirements by judicious location of industries.
* Introduction of discipline in water (surface and groundwater) abstraction and waste water discharge to effect water conservation.
* River flow regulation.

INSTITUTIONAL MECHANISM

Hitherto, water management in its various facets have been dealt

TABLE 3

Polluted River Stretches

River	*Polluted Stretch*	*Desired Class*	*Existing Class*	*River*	*Polluted Stretch*	*Desired Class*	*Existing Class*
1	*2*	*3*	*4*	*5*	*6*	*7*	*8*
Yamuna	Delhi-Itawah	C	D E	Sabar-mati	Sabarmati Ashram to Vautha	D	E
	City limits of Delhi, Mathura and Agra	B	D E				
Chambal	D/S Nagda and D/S of Kota (15 kms.)	C	D E	Satluj	D/s of Ludkhiana to Harike	C	D E
				Subarnar-ekha	Hatia Dam to Bharagora	C	D E
Kali	D/S Modinagar to confluence with Ganga	C	D E	Godavari	D/S of Nasik to Nanded	B	D E
Hindon	Saharanpur to confluence with Yamuna	D	E		City limits of Nasik and Nanded		
					Mancherial and Ramagundam to Bhadrachalam	C	D
Khan	City limit of Indore	C	E		Karad to Sangi	C	D E
	D/S of Indore	C	E		Dhom Dam to Narasarobadi	C	D
Kshipra	City limits of Ujjain	B	E				
	D/S of Ujjain	C	E				

Damodar	D/S of Dhanbad to Haldia	C	D E		Tributary Stream Nira	*	*
Gomti	Lucknow to confluence with Ganga	C	D E	Krishna	Upto Nagarjun Sagar Dam and from that Dam to up stream of Repella	C	D
Cauvery	Talakkavari to 5 km. of Musore Distt. Border Yagachi	A	C	Bhadra	Origin to D/S of KIOCL of Bhadra Dam (Karnataka)	C	E
	KR Sagar Ddam to Hogenekkall	C	E	Brahmani	Angul downstream from Kamalanga upto Bhuban	C	D
	Pugalur to Grnad Anicut	C	E				
	Grand Anicut to Kumbakonam	C	E	Tunga	Tirthahalli to Confluence with Bhadra	B	C
Kolong	Along Nagaon Town (Assam)	*	*	Narmada	Along Jabalpur city	B	B
Bhogdoi	Along Jorhat town (Assam)	C	D	Tapi	D/S Nepanagar to Burhanpur city	A	E
Barak	Along Silchar Town (Assam)	*	*	Betwa	Along Mandideep and Vidisha	C	D
Imphal	Along Imphal City (Manipur)	C	D	Beas	Upstream of Manali to Mandi	A	C
					Mandi downstream to Himachal Pradesh Border	A	C
Sowrah	Along Agartala City (Tripura)	*	*				
Sabarmati	Immediate upstream of Ahmedabad city to Ashram Sabarmati	B	E				

with in a compartmentalised manner. For management of water resources, both in terms of quantity and quality, it is necessary to have an integrated institutional mechanism. This is particularly needed in view of the fact that water resources transcend the political boundaries and the management of water cannot be and should not be addressed merely as a 'State' subject.

9

Fighting Forest Fires

Uma Shanker Singh

India has 7,65,210 sq. km. of forests of which 6,39,600 sq. km. have some kind of forest cover. India's forests are endowed with a great variety of biodiversity. The forests have been depleted by increased demand for fuelwood, fodder, timber, inadequacy in forest protection measures and diversion of forest lands for non-forestry purposes. Forest fires play a major role in degradation of biodiversity and productivity of forests. It is estimated that about 60 per cent of forest area is damaged by repeated annual forest fires. Forest fires are quite common and widespread since time immemorial. Mention of it is also found in the great epics like Mahabharata and Ramayana. Today, the incidence of forest fires is very high due to increased biotic interference in the forest eco-system and inadequate forest fire management practices. Decreasing fiscal allocation in forestry sector and change in our priorities towards the whole concept of systematic forest fire management are among the more important factors for failure in prevention of forest fires.

More than 92 per cent of forests in India are Government owned and the responsibility of forest management, including that of

* This article first appeared in Yojana, Volume 48 : 8 August 1997. Reprinted with permission.

preventing forest fires, lies with the state forest departments. Therefore, a national plan for systematic forest fire management for the purposes of guiding, coordinating and monitoring the activities in this regard is necessary to realise national goals and objective. The National Forest Policy 1988, also clearly spelt out the need for protecting forest fires through "special precautions to be taken during the fire season and improved and modern management practices adopted to deal with forest fires". But the fact remains that it is yet to get the priority consideration in forest planning and management it deserved.

Amongst various factors, which cause forest fires, the following are important:

- (i) Fires are caused deliberately by labourers collecting non-timber forest produce (NTFP), such as honey, sal seeds, etc.
- (ii) Fire left smouldering by the negligence of labourers/pedestrains/graziers, etc.
- (iii) By way of vengeance of villagers.
- (iv) Deliberately setting forests on fire by villagers with a view to getting fresh blade of grass for their animals as fodder.

EVALUATION OF LOSSES

This is one of the emerging areas in the forestry science which needs a comprehensive study, both area-wise and species wise. This kind of study will not only assess and reveal the loss of forest wealth/biological diversity but has the potential to direct the future forestry funding in the country. If the damages are quantified in fiscal terms, the value of the forest wealth would be properly appreciated. In India, there are two premier forestry institutions, viz. Indian Council of Forestry Research and Education and Forest Survey of India, with a clear mandate of carrying out comprehensive research on this subject. But none of these seems to have begun as yet with this type of studies. One of the studies conducted by U.P. Forest Working Plan circle puts the losses in reserve forest area on account of forest fires, at around Rs. 9 crore per annum with half of it, i.e. Rs. 4.5 crore, coming from the Hill Reserve Forest of Kumaon and Garhwal alone. It may be difficult to believe the veracity of methodologies adopted for quantifying these losses in monetary terms. Even if this loss is projected at the present rate, which is about 2.5 times, the loss works out to be Rs. 22-23 crore per annum. The indirect losses of forest fires in terms of loss in soil fertility, loss due to soil erosion, loss of annual increment to trees, loss of slow moving fauna, loss of natural regeneration of many species, etc. have not been taken into account while summing up the "net loss". If all those components are taken together for evaluating the "net loss", it would come to an astronomical figure. Besides,

TABLE 1

Forest Fire Incidences During 1991-94

Sl. No.	*State/UT*	*1991-92*	*1992-93*	*1993-94*	*Number of major fires*
1.	Andhra Pradesh	Nil	Nil	Nil	Nil
2.	Arunachal Pradesh	1	2	2	3
3.	Assam	Nil	Nil	Nil	Nil
4.	Bihar	7	15	10	Nil
5.	Goa	4	Nil	Nil	Nil
6.	Gujarat	507	633	654	Nil
7.	Haryana	Nil	Nil	Nil	Nil
8.	H.P.	605	352	600	79
9.	J&K	180	198	418	—
10.	Karnataka	106	16	Nil	1
11.	Kerala	211	90	112	Nil
12.	Maharashtra	1456	1428	Nil	Nil
13.	M.P.	629	371	461	34
14.	Manipur	2	4	6	Nil
15.	Mizoram	Nil	Nil	Nil	Nil
16.	Punjab	15	31	107	15
17.	Tamil Nadu	101	93	90	2
18.	Tripura	Nil	Nil	Nil	Nil
19.	U.P.	602	482	258	Nil
20.	A&N Islands	Nil	Nil	Nil	Nil
21.	Chandigarh	Nil	Nil	Nil	Nil
22.	D&N Haveli	13	33	23	Nil
23.	Daman & Diu	Nil	Nil	Nil	Nil
24.	Lakshadweep	Nil	Nil	Nil	Nil
25.	Pondicherry	Nil	Nil	Nil	Nil
26.	Delhi	1	1	1	Nil
	Total	4440	3749	1500	21

Source: Ministry of Environment and Forests.

TABLE 2

Areas Affected by Forest Fire

(Area in sq. km.)

Year	*Kumaon Division*	*Garhwal Division*	*Total*
1984	460	662	1122
1985	50	664	714
1986	22	05	027
1987	102	64	166
1988	314	49	363
1989	03	187	190
1990	129	05	134
1991	135	142	277
1992	208	151	359
1993	07	11	18
1994	54	06	60
1995	334	597	931

Source: U.P. Forest Fire Project.

it is not yet known how many flora and fauna in the forest ecosystem have been lost for ever on account of perpetual forest fires, which otherwise might have played an important and crucial role in the evolutionary process. Therefore, it becomes all the more important to evolve a methodology to assess the losses in real terms and not quantify them in plain financial terms.

Fire in the forests has been a perpetual problem though never tackled seriously; always a kind of "ad-hocism" has been followed. Even today, there is no co-ordination between Forest Department on one hand and National Remote Sensing Agency (NRSA), India Meterological Department (IMD) and Forest Survey of India (FSI) on the other hand. The department does not have the requisite resources to fight the fires. Therefore, there is not only an urgent need for a radical change in the curriculum of forestry programmes but also creation of a National Forest Fires Research Institute (NFFRI) to carry out coordinated and integrated research work on the forest fires, so that its implementation at the ground level is facilitated.

Information regarding forest fires generates from the unit level of forest administration, i.e. forest guard. A forest guard, in all probability, puts in wrong figures of fire-affected area in his weekly return and submits it to his superiors and this keeps on multiplying as it reaches the state headquarters level. It results in a much bigger numerical error at the national level when compiled together. It is not that forest guards like to give wrong information to their superiors but it is only because of the apprehension that either reporting a high rate of fire occurrence or reporting a larger fire affected area might land them in a tight spot resulting in "action" against them. This view also gets reflected in the following table, which shows the incidences of fires in the country.

TABLE 3

Project in U.P. (Haldwani, Nainital, Pithoragarh)

Total Project Area = 3,72,693 ha.

Year	*Percentage of area burnt (Of Total Area)*
1984	4.24%
1987	1.26%
1988	1.31%
1989	2.90%
1990	0.43%

Project in Maharashtra (Chandrapur)

Total Project Area = 1,62,844 ha.

Year	*Percentage of area burnt (Of total Area)*
1984	14.82%
1986	9.12%
1987	3.76%
1988	2.06%
1989	1.86%
1990	1.00%

Source: Ministry of Environment and Forests.

Table 1 shows that except Madhya Pradesh, Karnataka, Punjab, Tamil Nadu, Himachal Pradesh, and Arunachal Pradesh, no other States had any major forest fires and some States like Andhra Pradesh, Assam, Tripura had no forest fires at all. No forester will believe the theory that there would be no fire at all for three consecutive years. This also shows that how mortally afraid foresters are in accepting the truth and this not only results in poor budgetary allocation to the forestry sector but also drawing flak from others. If incidence of forest fires is reviewed year-wise in Uttar Pradesh alone it may be found out that every fourth year is a major fire year. Table 3 gives the forest fire incidences in Uttar Pradesh:

TABLE 4

Outlay Detail of CS and CSS During VIII Plan

(Rs. in lakh)

Year	*Outlay*	*Outlay/Expenditure*		
		CS	*CSS*	*Total*
1992-93	100.00	56.08	74.28	130.36
1993-94	100.00	54.60	123.88	178.48
1994-95	100.00	45.00	255.00	300.00
1995-96	100.00	55.00	345.00	400.00
1996-97	100.00	75.00	525.00	600.00
Total:	500.00	285.68	1323.16	1608.84

Source: Ministry of Environment and Forests (Annual Plan).

PREVENTION AND CONTROL

The following measures are adopted in preventing and controlling forest fires:

(a) Prevention

This is a pre-forest fire stage, which includes following components.

(i) control burning

(ii) clearing of forest lines

(b) Control

This is a post-forest fire stage which begins only after the fire has set in. It contains following components:

(i) Detection
(ii) Information
(iii) Fire-fighting

TABLE 5

Budget Allotment in Control and Prevention of Forest Fire in U.P.

(Rs. in lakh)

Year	*Plan establishment & machinery*	*Works component*
1	*2*	*3*
1985-86	11.60	13.06
1986-87	19.96	81.45
1987-88	26.43	76.98
1988-89	32.83	53.62
1989-90	43.72	42.83
1990-91	6.58	58.96
1991-92	33.42	18.15
1992-93	34.88	16.54
1993-94	36.49	11.06
1994-95	38.72	10.92

Source: U.P. Forest Fire Project

PRESENT STATUS

Though forest fire is a global phenomenon, the only difference in respect of tackling this problem in India is the lack of prioritisation of the forest management objectives. The concept of forestry planning has changed over the period of time with the changing fund planning. Earlier, most of the forest activities were funded through non-plan and, therefore, the funding source was stable and assured, but soon after the introduction of plan funding in the forestry sector, the maintenance components of forests, which included many issues (fire protection being one of them), were given up. Again, with the various social sectors competing with each other in successive plans, the funding in forestry gradually dwindled down with slight variations here and there. In the past, the control burning and clearing of forest lines were so important and sanctimonious to a forester that the time schedule fixed for these operations never even wavered. With the passage of time phrases like "Employment Generation" and "labour intensive sectors" etc. led to over-employment of people in the forestry sector and almost all of the plan funds were either distributed in their wages or in other establishment expenses. It was natural that these problems will come up with the increasing population, but it should have been decided by the overriding "principle of carrying capacity" or sustainable development of an organisation. Today it has ramified into a very different kind of socio-economic problem as the money which should have otherwise gone to work component goes elsewhere.

CENTRAL SCHEME

The importance of forest fires was felt at the Central level in 1984 when a UNDP Project titled "Introduction of Modern Forest Fire Control Methods" was introduced in selected areas of Uttar Pradesh and Maharashtra. The project was primarily to devise, test and demonstrate the principle and techniques of prevention, detection and suppression of forest fires. The project continued till 1990-91 as a pilot project and thereafter the UNDP assistance was withdrawn. The total UNDP input over the project period, i.e. 1984-90 was US $5,106,567 with the counterpart funding from Government of India at the level of Rs. 89.94 lakh. The results of the two project areas, namely, Uttar Pradesh (Haldwani, Pithoragrah and Nainital) and Maharashtra (Chandrapur) were impressive in the earlier stages, as shown in Table 3.

TABLE 6

Outlays in Forest Protection/Forest Fire in States

(Rs. in lakh)

States	*Name of the scheme*	*VIII plan outlay*	*1993-94*	*1994-95*	*1995-96*
1	*2*	*3*	*4*	*5*	*6*
Assam	F.P*	1035	358.75	441.00	427
Andhra Pradesh	F.P.	80	—	25.00	25.00
Bihar	FiP**	500	—	—	30.00
Chandigarh	F.P.	50.00	—	8.50	2.40
Dadra & Nagar Havelli	F.P.	—	—	5.00	6.00
Delhi	—	—	—	—	—
Karnataka	F.P.				120.00
	FiP	—	—	—	26.24
Kerala	F.P.	—	—	—	555.00
Jammu & Kashmir	F.P.	200.00	11.50	26.54	29.60
Himachal Pradesh	F.P.	50.00	10.35	25.00	20.00
Haryana	F.P.	15.00	4.00	4.00	4.50
Goa	F.P.	133.50	26.00	23.00	25.00
Lakshadweep	—		—	—	—
Meghalaya	F.P.	150.00	—	15.00	20.00
	FiP.		—		23.00
Pondicherry	F.P.		2.50	4.00	4.00
Panjab	F.P.	70.00	—	—	—
Orissa	F.P.		395.00	80.00	80.00
	FiP.		225.00		
Mizoram	F.P.	190.00	—	23.35	32.00
Tripura	F.P.	235.00	—	50.00	68.14
Tamil Nadu	F.P.	—	0.86	1.00	1.00
	FiP.	—	—	10.00	10.00
Sikkim	F.P.	17.00	—	5.00	20.00
Rajasthan	F.P.	340.00	—	15.00	55.00
West Bengal	F.P.	196.00	—	20.00	20.00
Uttar	F.P.	378.00	92.99	93.75	96.50
Pradesh	FiP.	75.00	9.00	10.00	10.00

* F.P. Forest Protection

** FiP Fire Protection

Source : State's Annual Plan.

Table 3 shows the importance of programmes, like forest fire protection and how they help prevent and check forest fires. However, after the withdrawal of the UNDP assisted project in 1990-91, Ministry of Environment and Forests (MOEF) did not examine the need to expand the scheme of prevention and control of forest fires all over the country in spite of the concern shown at various levels. Here it would be proper to mention that this programme initially started off at two places in two States only and later the scope of this scheme enlarged to five States.

SHORTCOMINGS

The scheme of introduction of modern forest fire control methods has two components, namely, maintenance of two helicopters (which had come from UNDP assistance) in Central Sector scheme (CS) and work component of forest fire protection to be taken care of under Centrally Sponsored Scheme (CSS). Table 4 shows the outlay details.

If the outlays in Table 4 are analysed, it may be found that Rs. 285.68 lakh have been spent on two helicopters. As the mandate of the Ministry of Environment and Forests (MOEF) goes, the two helicopters are supposed to be pressed into fire-fighting operations and inspection of forest and plantation areas. It has been found through experience that helicopters are highly ineffective in fighting fires as these aggravated the problem more than solving it. Helicopters are fitted with a water tanker below and is used to sprinkle water on the forest areas, which are ablaze; but its operational cost, water availability in the summer, danger of flying low are issues which make their operation very difficult and ineffective in the Indian context. Therefore, if proper planning and adequate home work would have been done prior to pressing helicopters into this kind of operation, the total cost incurred annually on two helicopters, which is roughly 29.19 per cent of the total, could have been utilised more effectively elsewhere like procuring vehicles for the rapid and immediate mobility of the fire crew from the range site to the fire site, wireless network to enhance rapid detection and other fire fighting tools. However, the helicopters have been invariably used in VIPs/VVIPs transport instead of their assigned work. With the kind of forest fire in 1995-96, on the hills of U.P. and H.P., MOEF should not take any chance now and work out a comprehensive action plan without losing any more time.

DECENTRALISATION

In hilly forest, areas the effective way to check forest fire expansion are maintenance of fire lines, compartment/block lines, installation of adequate watch towers, networking through wireless communication system and a highly mobile fire fighting crew to reach the nearest site of fire by road. All these facilities should be given at the level of Divisional Forest Officer (DFO) and not at level of Conservator of Forests, as in case of U.P. where fire fighting circle headquarters is at Haldwani. In this situation, if fire erupts at Almora, Pithoragarh or Nainital, fast deployment of all support system to fight forest fire can never be ensured immediately and effectively.

If the work component of the forest fire scheme in U.P. is analysed, it may be found that as the project progressed, its wage/ establishment component grew more than the work component as Table 5 shows.

Normally external assistance in forestry and wildlife sector leave behind withdrawal symptoms as soon as either it is over or withdrawn. The classic example of this can be seen in case of U.P. where soon after the UNDP withdrawal from the project, the persons employed in the project as daily wagers became jobless overnight. The budgetary in-flow in this programme from Government of India (GOI) as well as State Government of U.P. also dried up. Daily wagers approached honourable courts and brought a stay against their dismissal on the pleas that they had worked for years together without any interruption. All this led to sit ins and strike in such a magnitude that the work remained paralysed at Haldwani fire fighting headquarters for months together. Secondly, the assets which were created during the project started withering for want of proper maintenance.

The fire protection situation all over the country is precarious. It would be interesting to know that most of the state governments have not had any regular scheme for preventing and controlling forest fire. Table 6 gives the outlay in the scheme of forest fire and forest protection schemes, presumably understanding that forest fire may also be taken care of in the scheme of forest protection.

In the Earth Summit held at Rio in 1992, the problems caused by the forest fire in various dimensions of forestry were also discussed at length and these also got reflected in the para 11.2 of Agenda 21 as a basis for concerted action. It reads as follows:

"Forests worldwide have been and are being threatened by

uncontrolled degradation and conversion to other types of land uses, influenced by increasing human needs; agricultural expansion; and environmentally harmful mismanagement, including, for example, lack of adequate forest-fire control and anti-poaching measures, unsustainable commercial logginig, overgrazing and unregulated browsing, harmful effects of air-borne pollutants, economic incentives and other measures taken by other sectors of the economy. The impact of loss and degradation of forests are in the form of soil erosion; loss of biological diversity; damage to wildlife habitats and degradation of watershed areas, deterioration of the quality of life and reduction of the options for developments".

VARIOUS REPORTS

The importance of forest fire protection on the entire eco-system has been felt by all and it also gets reflected in various committee reports.

International consultants team of FAO gave a detailed technical report underlining the grim situation of forest fire in India. The report points out:

"Fire situation in India is alarming. There is no focus on forest fires in India in economic or ecological terms, and the curriculums of foresters' training at all level are rather inadequate in fire science and technology. There is no clear fire strategy nor an awareness of strategic planning. Fire management is only now beginning to engage the attention of planners and the awareness of fire effects on the forest economy is severely limited. Statistical data on forest fires, where available, are either skeletal or unreliable, as these data are not the output of a system of recording. No system exists for fire weather forecasting, danger rating, fire reporting or preventive measures apart from some basic fire-line clearance and prescribed burning on plantation boundaries. India's forests deserve a full-fledged fire management system institutionalized at the state level with strategic inputs of training, research, and awareness building coming from the Central Government."

Another committee, headed by Shri R.P. Khosla which was constituted by Ministry of Environment and Forests after a massive fire in the U.P. Hills and H.P. in 1995 suggested a number of remedial measures for fire protection. Some of these were:

(i) Increased vigilance is necessary by appointment of an adequate number of fire watchers during the month of April, May and June as used to be the practice earlier.

(ii) Clearing and maintenance of fire lines virtually abandoned of late due to shortage of funds must be carried out regularly.

(iii) The practice of control burning to deal with the accumulation of combustible pine needles on the forest floor be reintroduced.

(iv) Efforts for finding alternative uses for pine needles should be supported by the government so as to demonstrate their economic viability. This will help reduce the accumulation of combustible material on the forest floor.

(v) The forest department staff should be provided with a complete communication network through wireless to enable a quick response in dealing with forest fires and also with the problems of illicit felling of trees.

(vi) The communication network should be supported with improved mobility to enable quick transport of men and materials from one area to another. For this at least one additional jeep may be provided at the divisional level to the D.F.O. in the hill area.

(vii) Whether villagers come to assist the forest department in extinguishing forest fires or not, their timber rights should not be curtailed or forfeited.

(viii) The state government must ensure that adequate funds are provided to the Forest Department for the proper care, maintenance and protection of the forest. These funds should be provided through a centrally sponsored scheme for this purpose.

10

Joint Forest Management: Case Studies

Amit Mitra

Marking a sharp departure from the past, 15,000 community groups, across 16 states in the country, are managing 1.5 million hectares of forests jointly with the forest departments. This has been possible under the Joint Forest Management Programme (JFM), widely acclaimed as the first step towards an era of pro-people forest management.

The evidence coming in from these states, however, after seven years of JFM, seems to affirm that it has been relegated to just another programme. Rhetorically based on concepts like 'participation', 'sustainable development', it has hardly empowered people to manage their own resources. Is this because there is something fundamentally wrong with the basic concept itself or is it a question of improper implementation? Can the situation be redeemed? Protagonists of the JFM approach argue that it is a process and it takes time to change attitudes that have been formed over a century and a half. Opponents say that it is a mere eyewash. Introduced under international donor agency pressure, it seeks to appropriate people's movements to

Reprinted with permission from Yojana, August 1997.

regenerate forests and buy protection cheaply.

This article examines the concept and actual practice of JFM. Apart from the secondary literature, it is based on field observations, between November 1996 and May 1997 in Karnataka and Orissa.

GENESIS

JFM is a direct outcome of the 1988 National Forest Policy, which acknowledged the dependence of the rural poor on forest resources for survival. This policy extolled the virtues of people's participation in development and conservation of forest lands. For the last 150 years, 23 percent of the geographical area of the country (329 million hectares), which has been classified as forest lands, has been government-controlled, under the aegis of "planned and scientific" forestry and has focussed on timber production and extraction to fulfil national needs. Human made plantations of industrial species were encouraged on forest lands, often by clear-felling mixed forests. Even after independence, this policy continued (Saxena 1995: ix).

This led to deforestation, accentuated rural poverty and unemployment. Social unrest and conflict over rights increased, due to the alienation of the local community groups from the resources on which they had depended for centuries. Degradation continued unabated and the situation was compounded by the development of forest based industries and increasing biotic pressure. Scarcity of agricultural land (2.623 million ha of forest land was officially transferred to agriculture between 1951 and 1980), encroachment, transfer of land for development activities and industries enjoying heavily subsidised flow of raw materials from the state forests continued to accelerate the degradation process at the cost of the rural poor (Agarwal and Sehgal, 1996:1).

The government did try to curb the depletion of forests by putting it in the concurrent list under the jurisdiction of both the Centre and the State in 1976. The Central Government acquired the powers to pass the laws concerning forests and wildlife. A major step constituted the passing of the Forest Conservation Act in 1980, which made the Central Government's approval mandatory for diversion of forest land for non-forest use. The diversion of forestry lands to non-forestry use was checked somewhat, but forest degradation continued, at the rate of 47,300 hectares annually between 1980 and 1985 (Agarwal and Sehgal, 1996: 1).

To ease the pressure on forests, a social forestry programme was launched on non-forest land to meet people's subsistance needs but it failed to yield the desired results. Social forestry failed to check the relentless battle between foresters and the users of forest produce. Many foresters began to question the methods and practices of the forest departments. They started to realise that involvement of the people is a must for the success of any forestry programme (Agarwal and Sehgal, 1996: 2).

At the same time, information began to materialise about many village groups protecting forests adjoining their villages on their own. There are many thousand such informal forest protection groups in Orissa, south Bihar, Madhya Pradesh and Gujarat. Some far sighted forest officials in West Bengal started involving the fringe communities in forest management. The results were dramatic in terms of forest management.

It was in this context that the National Forest Policy of 1988 was formulated. This involved a reorientation towards prioritising environmental concerns as well as meeting local needs and involving people in forest management. According to this policy, forests are not to be commercially exploited for industries, but these are to conserve soil and environment and meet the subsistence requirements of local people. The policy gives higher priority to environmental stability than to earning revenue (para 2.2) It discourages monocultures and prefers mixed forests. The focus has shifted from commerce and investment to ecology and satisfying the minimum needs of the people, providing fuelwood and fodder and strengthening tribal forest linkages (Saxena, 1995: 13). Para 4.3 of the new Policy reads as follows:

The life of tribals and other poor people living within and near forest revolves around forests. The rights and concessions enjoyed by them should be fully protected. Their domestic requirements of fuelwood, fodder, minor forest produce, and construction timber should be the first charge on forest produce.

However, inspite of the forceful language of need based, poverty focused forest management, there was little of a concrete nature to operationalise these objectives.

OPERATIONALISING NFP

Ultimately, the 1988 policy was sought to be implemented through a June 1990 Ministry of Environment and Forests (MoEF), Government of India circular, giving birth to the JFM concept. The circular is to a large extent based upon the successful experience of joint management in West Bengal. It recommends that village institutions constituted specifically for forest regeneration be given usufruct rights and a share of revenues in delineated forest areas (MoEF, 1990, paras 4.5). The rights and obligations of beneficiaries are to be drawn up as a Memorandum of Understanding (MoU) and agreed by both participants and the State Government (MoEF, 1990, paras 1.7.14). Moreover, the 1990 circular specifically recommends the involvement of NGOs in "building meaningful people's participation in protection and development of degraded forest lands" (Para 3) owing to their expertise in community organisation. The circular does not, however, make clear whose responsibility it is to develop village institutions, nor does it give such institutions any right to become involved in forest management.

As a national policy its recommendations are flexible and require the parallel formulation of a state policy and its operationalisation through government orders.

The JFM programme has been adopted by 16 states of the country by issuing resolutions. The only exceptions are Kerala and the states in the North-East. Kerala too seems to be working on its JFM resolution. In the North-East, the forests are mostly owned by the local communities and the JFM, as it is understood, in the rest of the country is not applicable in its present form. Tripura, however, has issued its own JFM resolution and progress has also been reported from Meghalaya and Arunachal Pradesh.

As the forest types and socio-economic parameters vary greatly across the states, the JFM orders too differ in terms of composition of the forest management group, its type (society, cooperative, FD registered), usufructuary and timber benefits and so on. The thrust of the JFM orders, in the states, however, is similar. In the introductory paragraph of all the orders, the need for community participation is stressed to promote regeneration and development of degraded forests. The Orissa Order (1993) goes even further to acknowledge that forest management has to be reoriented to forge an effective parternship between the forest department and village communities and that equal partnership is envisaged for the two.

JFM has won a lot of support even outside the forest department's circle, because it is politically correct and places people at the heart of forest management. It is held to represent a fundamental shift in forest regeneration methods and conceptually envisages a movement.

From	*To*
Centralised management	Decentralised management
Revenue orientation	Resource orientation
Production motives	Sustainability
Single products	Multiple products
Large working plans	Microplanning
Target orientation	Process orientation
Unilateral decision making	Participatory decision making
Punitive rules	Self abnegation rules
Controlling people	Facilitating people
Department	People's institutions
Assumed homogeneity	Recognising diversity
Achieving pre-set objective	Fulfiling multiple, need based objectives
Area management	Site specific management
Timber production	Multiple products combined with biodiversity
Single technical package	Menus of options
Fixed procedures	Experimentation and flexibility
Plantation as first option	Low input management and regeneration
Single species	Multi species and multi-tier plantations

Source: Raju, 1996.

POOR PRACTICE

The practice, as the evidence from all across the country shows, leaves very much to be desired. This is notwithstanding the success stories; like Arabari in Midnapore district of West Bengal, which have demonstrated beyond doubt that local communities can protect forest patches near their villages and that the forest department too can work with the people if it wants. The basic principle of JFM, as articulated by the forest department is that:

- Large areas of degraded forests can be regenerated by involving the local community in protection.
- The local community will protect and regenerate the forest if they have the incentive and power to exclude outsiders.

The practice, however, shows that:

- In most areas, JFM is being used as a strategy to regenerate degraded forests and improve survival in plantations to meet departmental objectives.
- While JFM orders allow free access to non-timber forest produce (NTFP)—except those that are nationalised—the management objectives are departmentally set and mostly timber oriented.
- The communities are not so much interested in the benefits from a final harvest, as emphasised in the JFM framework as they are in a continuous stream of benefits from the forest. This includes from the forest. This includes NTFPs and fuelwood and fodder. Indeed, all the malaise of typical target oriented government programmes mark the implementation of JFM. This extends even to the formation of village groups, called by various names like Van Surakshya Samities (Orissa) or Village Forest Committees (VFCs) across the states. Prior to the visits of VIPs or review teams of bilateral donors, senior staff like the conservator of forests, are seen to rush around the countryside forming the village committees.

Or take for instance the microplans, which is an integral part of JFM. All the state government orders on JFM envisage the preparation of a microplan with the participation of the village institutions. The microplans are supposed to reflect the people's needs and the silvicultural practices changed accordingly. In practice, the state forest departments tend to dominate the preparatory process and the final document reflects more of the investment to be made rather than the benefits or the resource use profile by the community. The forest departments usually fight shy of parting with information regarding budget, wage and piece rates that are due to the labour. The forest departments continue in their own way of afforestation, although on paper the microplan becomes the basis of investment. Indeed, the micro-planning exercise becomes a source of manipulation, as the experience from across the country shows. In Taladandakla village of Phulbani district or in Kutling village of Mayurbhanj district in Orissa, the villagers had not even heard of the microplan, though it was supposed to have been prepared jointly with them. In Karnataka, the villagers are supposed to indicate their choice of trees to be planted and this is recorded in the microplan. In Hitlali village, Uttara Kannada district, again, the villagers, had not heard of the microplan. At a much later date, after the plan had been prepared and the pantation work completed, an arbitrary list of the trees that the villagers 'chose' was planted in the microplan. In almost all the states, the microplans are written in English, a language that the villagers generally cannot read, even if they get access to the document. Nor is the manipulation restricted to the miocroplan alone. A crucial document in JFM is the signing of the memorandum of understanding with the village institution. In Kangod village of Uttara Kannada district in Karnataka, the first village in the state where an MoU was signed, the village institution's president was made to sign on a blank MoU because a minister was coming for the inauguration and there was no time to go through the process of explaining things to the villagers. In Gujarat, the "forest department, while carrying out investment plans treats the village institution as a body of individual labourers and pays them their wages for the works carried out. This is contrary to the spirit of JFM where the forest department is supposed to interact with the community as a collective, that is interact with the village institution as a whole and not as individuals". (Raju, 1997: 11). Sometimes, there is open animosity, as in Bellankeri, and Bashi villages in Uttara Kannada district of Karnataka, where marginal farmers have lost all their lands for encroaching an acacia has been planted on those very plots under JFM. The people are reduced to mere numbers in a target chasing game, with hardly any say in the actual management of the forests.

REVENUE MAXIMISATION

The orientation of the forest department continues to be heavily oriented towards revenue maximisation by generating timber. This gets reflected in the emphasis on monocultural plantations, even on lands that can be regenerated naturally. Natural regeneration is possible in most parts of the country and allows the growth of a multi-species forest, which can meet the diverse needs of the community. The only problem with natural regeneration is that it takes time for the forest to grow and the departmental target of plantations cannot be met. But since a teak or an acacia plantation does not meet the survival needs of the local people, they do not participate in such programmes.

Moreover, JFM is restricted to degraded areas, especially those with tree cover of less than 25 percent. This is in gross violation of the 1988 national forest policy. In Uttara Kannada district, for instance, this restricts people's participation to just 1 per cent of 8262 sq. km. the area of the district classified as forests. The dergraded areas are often used as pastures by the communities and undertaking monocultural plantations on them means a net loss to the community.

The JFM orders stress on the sharing of a final harvest. However, the communities are more interested in the usufructs and not the final benefits. NTFPs, for instance, mean a lot more to a poor household than a share in the timber 15-20 years later. In Orissa, almost all villages protecting their forests aver that they are not doing it to fell the trees and getting a final harvest. Rather, the NTFPs, the leaf manure and soil conservation is more important. But in Karnataka, the government order (GO) specifies that the people purchase the NTFPs from the government depots, so that they do not fall in the clutches of the middlemen. So the people have to pay for what they protect!

There are many more such examples which violate the basic philosophy of JFM. Indeed, it is difficult to find a JFM village in most of the states where the people consider the initiative and the institution as their own and not a government programme.

In practice, however, the attitude of the forest department remains what it was. This is exemplified in the Orissa case, where communities have been protecting and regenerating forests for years, without any outside help. The Government of Orissa has been negotiating for money with the Swedish Development Agency (SIDA). To prove its 'success' it has appropriated many of the indigenous efforts, often with false promises. In one village, Karlamal, in Phulbani district, the villagers consent was got by free distribution of T-shirts and giving a colour television set to the village. A radical government order was issued on 30th September, 1996, which declared that the forests being protected by the community would be declared as village forests. Today everyone, starting from the PCCF to the DFOs, try to disown it by saying it cannot be implemented. Why pass a GO which cannot be implemented?

STATUTORY BACKING

This brings in a related dimension. None of the JFM resolutions or orders have any legislative backing. There have been repeated demands from the field for this, but to no avail so far. Essentially, the government and the forest department views participation, in terms of the inclusion of human resources, as the key input in development projects to ensure its success. The forest department is just interested in cheap forest protection and JFM is a way to achieve it. Participation is not seen as a process by which the ordinary people gain influence and control over their resource base and improve their lives. JFM is not seen as a method of empowerment.

The JFM practices are in stark contrast to the high level of identification that villagers have and the sense of empowerment they feel wherever they have taken the initiative to protect forests on their own, independent of JFM. There is now increasing evidence of such community protection intitiatives from Orissa and Bihar, where thousands of village groups are regenerating an estimated 400,000 ha of forests and to a smaller extent in Gujarat, Rajasthan, Karnataka, Madhya Pradesh and Andhra Pradesh (Agarwal and Sehgal, 1996). The Primary JFM principle, that the communities can manage their forests, is said to have been enunciated from these experiences.

These village groups have started protecting and regenerating forest patches adjoining their villages in response to the forest degradation and the ensuing fuel, fodder and minor forest produce crisis. These groups have mostly come up in the areas where people continue to have strong economic and cultural dependence on the forest and where traditions of community resource management still survive. These people initiated forest protection without any outside help, governmental or non-governmental, and have shown a tremendous capacity to manage the forests. Elaborate institutional mechanisms have been developed by these groups to share costs and benefits among all the constituent households. Mechanisms to control access and impose fines on free riders are decided in village meetings and consultations.

Though there are many variations and adaptations across the villages, the broad rules formulated by these indigenous community institutions are:

- Right to fuelwood collection. The villagers are permitted to collect dry twigs, fallen branches, but are not allowed to cut live trees. The variations that exist in this are: collection once a week or once or twice a year, but on specified days, when the forest is opened for the entire group, with the committee members supervising.

- Right to collect non-timber food products such as fruits, roots, leaves, gum, tasser, vegetables and oil seeds.
- The people apply to the managing committee for the wood they need for agricultural implements and house construction/repairs. The committee verifies the genuineness of the need and allocates some trees according to requirements. Sometimes a token fee is charged and the money is given to the village development fund. Often, there are norms to govern the frequency of the applications for wood and the species that can be so used are specified.
- No commercial felling of trees is permitted.
- Compulsory participation in guarding the forest, mainly by a rotational watch, where one/two members from specified households act as watchers. This is known as *thengapali* in Orissa or *vara* in Gujarat. Sometimes villagers pool grain to appoint guards.
- Fines to punish erring villagers and outsiders.

All these village institutions score very high on norms of TRAITS—transparency, rules, awareness, initiative and development, tenure and satisfaction of needs (Raju, 1996: 105). They need to be recognised and respected as independent people's institutions by both governments and the non-governmental organisations.

If the forest departments and the government really want to involve people in regenerating India's forests, it has to change the very basic farmework of JFM, in response to the field realities and the experience of community forest management. The new framework would entail first recognising the village forest committee as an independent entity and not an appendage of the department, which exists just to protect. It has to be given the authority and autonomy to manage the forest. This means that the forest department will not be central to JFM as it is today but the people will be given the primary importance with decision making powers. The focus of forestry will have to change from timber and commercial forestry to NTFP's and meeting daily needs of the people. Ultimately, the forest department will have to play the role of an adviser and facilitator like the agriculture or animal husbandry departments. That is the only way to save our forests and there is enough evidence, from isolated cases of individual officers, that the department can do so.

11

Wild Life Legislation and Administration

B. Bharatha Lakshmi

Prior to independence the protection of wild life and brids was the responsibility of local forest officers (under Indian Forest Act, 1927) and the rulers of the area. After independence till 1972, when the wild life (protection) Act, 1972 was enacted, no uniform national legislation was available to control and regulate the wild life activities, excepting a few voluntary conservation agencies like Bombay Natural History Society trying to create public awareness among the masses for protection of wild life. As a result most of the colourful wild fauna and flora are today threatened and endangered.

The Bombay Natural History Society, established in 1883, as a private agency took the first setp in formulating a comprehensive and detailed act for Bombay in 1951, which was known as Bombay wild animals and wild birds protection Act, 1951. Other special legislations enacted by different state governments were, the Elephant preservation Act. Rhinoceros preservation Act and the Games Act. There was a need to enact uniform and national legislation on wild life activities. Wild life control administration, management, National parks and sancturies were essential for protection, conservation and efficient management of the left over wild heritage of our country. With these objectives, the Wild Life (Protection) Act, 1972 was enacted and extended to in all states except Jammu & Kashmir and Nagaland.

Wild Life (Protection) Act, 1972 is the primary legal document for wild life management all over the country.

The Forest (Conservative) Act, 1980, prevents the denotification or use of any forest area for non-forestry purposes.

The Wild Life (Protection) Act, 1972 deels with protection of wild animals and birds and matters connected there with, ancillary or incidental.

It has 7 chapters, 66 sections and 6 schedules of animals. They deal with sanctuaries National parks, administration of wild life organisation regulation of hunting of wild animals trade in animals and animal articles, Prevention and detection of offences and miscellaneous.

Chapter (Sections 1 & 2) deals with priliminaries and definitions. Chapter 2 (Sections 3 to 8) deals with administration and appointment of officers such as Director of wild life preservation in the central Government, chief wild life warden and wild life wardens in the states. Chapter 3 deals with regulation of hunting of wild animals. This chapter 3 includes section 9 to 17. The wild animals birds, amphibians and reptiles are grouped into 6 different schedules.

Ist schedule includes all the endangered and rare species and their hunting is completly prohibited throughout the country eg. (siberian white crane (*Grus leucogeranus*). Brow antelered deer or Thamin (*Cervas eldi eldi*) peafowl (*pavo cristatus*), Musk deer (*Moschus moschiferus*) etc. Schedule II consists of special games. Schedule III includes bigger games like sambar (*Cervus unicolor*) Schedule IV compraises small games such as Duck, cranes, Bustards. The last schedules include the vermins such as common crow, mice, rats etc.

The schedule pattern has brought uniformity in the control and supervision throughout the country and the animals and birds are protected with equal measure everywhere. A notification in 1980, separates schedules drawn for the flora part of wild life. For ex *Dioscorea deltoidea* in I-B.

Chapter 4 of the act exclusively deals with creation, management and legal provision for the national parks, sanctuaries, reserves and closed areas. Now trade or commerce in wild animals or animal products has also been regulated through sections 39-49 of Chapter 5 of the Act. Penalties and Penal measures have been drawn for hunting the animals of schedule I and schedule II.

Thus Chapter 5 includes penalties and preventive measures for violation of the provisions of the Act. It also deels with the powers of the differs in preventing commitment of offences enforcement of penalties. Punishments, for violating the provisions of this act. It includes sections 50-58 of the act.

The Last Chapter 6 (sections 59-66) deals with miscellaneous aspects, like impact of the act on conservation.

The creation of schedule and strict provision for safeguarding these animals, resulted in the improvement in their population in states are killing of wild life was reduced and many rare and endangered species are saved from extinction. The trade in wild animals was brought under strict control and watch.

WILD LIFE ADMINISTRATION

Before the enactment of Wild Life (Protection) Act, 1972 the administration of wild life was the part of the normal duties of a forest offices and no separate organisation was specially created for it. But the Wild Life Protection Act, 1972 made elaborate and comprehensive provision for the administration of wild life. Both at central as well as state level, wild life wings have been carved out of the forest departments in all the states. At the central level, the Inspector-General of Forests (I.G.F.) is the highest exclusive authority dealing with forest and wild life in Ministry of Agriculture and he is assisted by one joint secretary for forest and wild life designated as Director of wild life preservation. Government of India. There are a Deputy Inspector General of Forests (D.I.G.F.) and Assistant Inspector General of Forests (A.I.G.F). There are 4 zonal assistant directors of wild life preservation, stationed at Delhi, Madras, Calcutta and Bombay.

All international trade in wild life and their products can only take place from these 4 parts of country and the species allowed to import or export are mentioned in the central export policy.

At the state government level chief wild life warden is the highest authority having statuary power under wild life Act, 1972. The rank and status of the chief wild life wardens is diffrente in different states. In some states the post is equivalent to chief conservator of Forests or Additional chief conservator of forests or a Conservator of Forests. At present forestry and wild life are integral part of the forest administration. The chief wild life warden enjoys statuary power under

Wild Life (Protection) Act, 1972. There are wild life wardens and Assistant wild life wardens who exercise their power as per the provision of the Act. In most of the states, the Divisional forest officer is designated s wild life warden.

Thus a line of administration set up has come up both at central as well as at state level, for guiding and monitoring the wild life management. In addition the zoological and botanical parks are also becoming popular in attracting the common public and managed by the state forests departments under the direction of wild life wing.

Indian board for wild life is the main body which advises the union government on all the wild life policy matters. The P.M. is its chairman. The members of the board include officials of the Government as well as eminent conversationists and naturalists. The board was first constituted in 1952 as was called as central board for wild life with 25 members. The board suggested measures for protection and safeguarding wild life and was later redesignated as Indian Board for Wild Life (IBWL). The membership of the board at present has gone up to 70. Main functions of IBWL are: (i) To advise the central and state Governments on the ways and means of conservation and control of wild life, through co-ordinated legislature and the declaration of certain species of animals as protected animals and the prevention of indiscriminate killing. (ii) To promote public interest in wild life and the need for its preservation, in harmony with the natural and human environment. (iii) To sponser the setting up of national parks, sanctuaries and zoological gardens. (iv) Advise the Government on policies regarding export of living animals, skins, furs feathers and other products of wild life. (v) to review from time to time, the progress of wild life conservation in the country and suggest such measures for improvement as are considered necessary. (vi) To perform other functions which are relevant to the purpose for which the IBWL is constituted. (vii) Other things.

The board from time to time also recommends amendments to the act and in shifting and adding any species of wild life in the Schedules of the act.

The IBWL through its recommendations and persuation has been able to create as many as 221 reserves which include 19 national parks, 202 sanctuaries for ensuring the protection of endangered and rare flora and fauna listed in schedule land 2 of the Wild Life Protection Act, 1972. The establishment of modern zoological park at New Delhi in 1955 was the result of the decision taken in its first meeting. Animals are given feeling of their natural habitats. Visitors can see the animals in open enclosures. On the recommendation of expert committee, central Government has started extending financial assistance through the five year plans for the creation and development of national parks and sanctuaries. When India became a party in the convention on international trade in endagered species of wild fauna and flora in 1976, the export and import of wild life has further been checked. The member countries by co-operating with each other have been able to bring the illicit trade to a minimum level.

The Inspector general of forests and director of wild life preservation are the management authorities. Directors, zoological survey of India Director, Botanical survey of India and the Director, Central marine fisheries institute have been declared scientific authorities.

State Governments have set up state wild life advisory boards for assisting the state Government in formulating board principles of wild life managements. These Boards provide expert recommendation after considering the local factors.

SOME WILD LIFE RESEARCH ORGANISATIONS

World Wild Life Fund (International)

It acts in close relationship with the international union for conservation of nature and natural resources IUCN (1961). It is a voluntary organisation with its headquarters at Glands (Switzerland). India has given financial and equipmental assistances to the following projects.

1. Project Tigher
2. Gir ecological research projects
3. Project Hangul
4. Himalayan Muskdeer Project.

Wild Life Preservation Society of India Established in 1958

Headquarters at Dehradun. Main object is to create public awarness for the preservation of wild life heritage of the country. The society attempts to achieve :

1. To promote interest in and impart knowledge regarding the preservation and conservation of all forms of wild life

vegetation, water and environment, particularly among the youth of the country.

2. To promote wild life tourism.
3. To carry out publicity and propaganda for the preservation of wild life in India by means of monographs, journals, bulletins, films, film strips and other feasible methods.
4. To assist the Government in enforcing the provisions of Indian Forest Act, and Wild Life (Protection) Act and helping in bringiing the culprits to books.
5. To advise and render help to the Government organisation for better management of National parks and Sanctuaries, preservation of wild life, raise species and in introduction of exotic species and finally rehabilitation of the rare species etc.
6. To cary out research work in matters of wild life management, rare and threatened species and related matters.

Bombay Natural History Society

B.N.H.S. is only one of its kind in the sub-continent. Since its foundation in 1883 as a private body at Bombay, it has vastly contributed in the field of research, conservation, education and disseminating and extending the knowledge of flora and fauna among all categories of population. The society is represented by Indian Board for wild life and state advisory boards of many states.

The Journal of the Society publishes articles contributed by the professional scientists, research workers, specialists and amateurs on various topics of natural history concerning with observation studies, survey and experiments, thereby making it a journal of very high international standard. The journal is one of the best sources of information and knowledge on the oriental flora and fauna through publication of special books such as Indian Birds by Salim Ali, book of Indian Animals by S.H. Pratar, some beautiful Indian trees by rev. E. Flatter, Game Birds by Baker, India's wild life by M. Krishnan, Birds of India and Pakistan by Salim Ali and S. Dillen Ripley etc. These publications helped in propagating the idea of conservation of flora and fauna and better understanding of nature and in dissiminating the knowledge of flora and fauna of the sub-continent. The Society has a library consisting of more than 6000 publications in various subjects of natural history.

It also assist in the form of grants in Aid for carrying out field work in natural history to students, specialists and naturalists from the fund created such as (Dorabji) Tata Trust, field work, fund, the Salim Ali-Loknantho orinothological Research Fund etc. The society in collaboration with foreign agencies carries out the research projects of Indian Mammals from 1911 to 1923. The orinothological Survy (in 1929) (Western Ghats) is still continuing. Large number of specimen were captured from all over India. The central crocodile breeding and management training, research institute has comeup with the assitance of food and agriculture organisation and UNDP-The institution impart training to forest department personnel in crocodile breeding management.

It is now heartening that the common man has realised the ethological and ecological significance of this beautiful wild life on the earth and that wild life conservation is imperative for the Doke of posterity as well.

References

Natraj Publishers, 1990. *The Indian Wildlife (Protection) Act*, 1972. Dehradun.
A. Krishan, 1996. A.P. Forest Laws.
Asia Law House, Hyderabad.
Law Publishers, Bare Act, 1997. The Wildlife Protection Act, 1972 with allied rules.
Natraj Publishers 1998: The Wildlife Protection Act, 1972 (Amanded upto 1991).
Asok Kumar, 1998: Wild Life (Protection) Act, 1972.
Ministry of Environment and Forests, Govt. of India.
Natraj Publishers in collaboration with Wild Life Protection Society of India 1998: Handbook of Environment, Forest and Wild Life Protection Laws in India.

12

Role of NGOs in Environmental Conservation

N.L.N.S. Prasad

India has been endowed with rich natural resources of both renewable and non-renewable. While exploitation of these resources is necessary to meet the growing needs of human population, their unplanned and unscientific exploitation would result in severe environmental problems. It is, therefore, necessary that a careful planning of utilisation of these resources is absolutely essential for sustainable development. As a developing country with a large population living below the poverty line, there have been always opposing factors between environment and development. Environmental degradation is a global phenomenon. About 70% of the World is witnessing the painful results of the deterioration of natural resources, particularly, the poorer nations, of which India is a part (MoEF, 1994). A number of measures have been initiated by the Government in the conservation of environment and also natural resources. The Government has enunciated Policy statements on forestry and on abatement of pollution. A National Conservation Strategy has also been formulated as part of the efforts in conserving natural resources. A number of laws have also been enacted such as the Wild Life (Protection) Act, 1972, the Forest (Conservation) Act, 1980 and the Environment (Protection) Act, 1986 as part of the measures to implement the policy to protect, preserve and promote environmental quality. Various rules have been made along with amendments in these Acts from time to time to make them more effective/stringent. A number of action plans and initiatives have also been made by the Government for environment protection and the conservation of natural resources, particularly, in the wake of the 1972 Stockholm Conference on Human Environment and subsequent Rio Summit in 1992. The experience for the last several decades provided a new approach that considers a grass-root level participation by the people at local as well as regional level to serve as a basis for successful implementation of various developmental programmes. Thus, the National Environment Action Programme conceived by the Government of India envisages an active participation of NGOs/Voluntary Agencies in creating awareness of environmental issues among the locals to ensure proper utilisation of local resources at the time of implementation of developmental projects so that the benefits reach the locals as well as make these projects sustainable (MoEF, 1994).

It is estimated that there are over 10,000 NGOs in India ranging from reputed agencies at national level to small local groups, from established research organisations to mass based field organisations. Many of these are engaged in popularising Eco-development Programmes, Waste Management, Forest Conservation, Preservation of Genetic Diversity, Environmental Protection, Bio-Agriculture, Eco-friendly Technologies in Industry and other environment related activities (Anon, 1995). A Directory was published on Environmental NGOs in India by the World Wide Fund for Nature (WWF)—India in 1994 which gives a detailed information of around 1500 NGOs engaged in various environmental activities and also health education and upliftment of rural poor. The Statewise list of NGOs is compiled and provided in Table 1. There are 102 NGOs in Orissa which have been registered under the Societies Registration Act of 1860. The list seems to be incomplete, as there may be more NGOs both registered and unregistered since the time of WWF Report. Majority of these organisations are engaged in eco-development activities, environmental awareness campaigns, water-shed management, forest protection, soil conservation, tree planting and fodder and fruit crops cultivation etc. However, there are very few organisations that are committed to research in the field of environmental protection and nature conservation in Orissa.

TABLE 1
Status of NGOs in India*

State	*Number of NGOs*
Andaman & Nicobar	02
Andhra Pradesh	169
Assam	34
Bihar	105
Chandigarh	01
Delhi	78
Goa	05
Gujarat	70
Haryana	14
Himachal Pradesh	17
Karnataka	56
Kerala	60
Madhya Pradesh	52
Maharashtra	103
Manipur	80
Meghalaya	03
Mizoram	01
Nagaland	06
Orissa	102
Pondicherry	04
Punjab	21
Rajasthan	68
Sikkim	03
Tamil Nadu	160
Tripura	04
Uttar Pradesh	159
West Bengal	112
Total	1469

* Compiled from the "Environmental NGOs in India—A Directory, 1994" by WWF—India.

This paper presents the natural resources and the developmental activities vis-a-vis environmental problems and the NGO's role in mitigating the impacts with special reference to Orissa State.

1. NATURAL RESOURCES OF ORISSA STATE

Orissa has a geographical area of 1,55,707 sq. km. contributing to 4.7% of total geographical area of the country. As per the 1995 satellite imagery data, it was estimated that the actual forest cover was 47,107 sq. km. which is about 30.3% of the State's geographical area. As per the 1997 assessment, the forest cover declined to 46,941 sq. km. (MoEF, 1998). As part of the National Wildlife Conservation Strategy, a number of steps have been initiated to protect forests as well as wildlife of the State. According the booklet Wildlife of Orissa—At a Glance brought out by the State Forest Department, Government of Orissa (1997), there are 18 Wildlife Sanctuaries and 2 National Parks in the State. These protected areas constitute 10.37% of the total forest area in the State. Two National Parks and 17 more wildlife sanctuaries have been proposed to be created in the State. Orissa State also has a varied geographical conditions from coastal sand beaches to high mountain areas of Koraput, Keonjhar and Mayurbhanj districts which form part of the Eastern Ghats. There are also a number of major rivers flowing through the State which include Mahanadi, Subernarekha, Brahmani, Baitarani. Besides, a number of tributaries such as Indravati, Rushikulya, Chitrotpala, Daya, Bhargabi, Devi and Kathjori also exist. Apart from the rich forest wildlife resources, there are a number of mineral deposits in the State which include iron ore, limestone, dolomite, bauxite, manganese quartz and graphite. In addition, rare earth elements are also available along the sand beaches of Ganjam District. A number of developmental projects with respect to mining, industries, river valley projects, thermal power plants and coastal harbour projects have been in different stages of implementation in the State.

Most of the river valley and mining projects are located either in forest areas or very close by them. These projects involve diversion of forest area and rehabilitation of affected people. Other environmental considerations include catchment treatment to prevent soil erosion and subsequent siltation of reservoirs, protection and restoration of construction areas in the case of river valley projects. The mining projects need to implement reclamation of mined areas, overburden dumps through construction of garland drains and plantation as well as prevention/mitigation of air, water and noise pollution. The actions of the project developers as well as the Government at Central and State level are directed to mitigate various environmental impacts. The role of Non-Government Agencies (NGOs)/Voluntary Agencies (VAs) to achieve a better quality environment is more pertinent at site level by involving the locals. Even other developmental projects require the cooperation and active participation of the NGOs to work for the involvement of the local people in their successful implementation and in the environmental protection. Therefore, the involvement of NGOs to work at grass-root level in the creation of environmental awareness as well as various issues and problems of developmental activities has been recognised by the Ministry of Environment & Forests. Even the World Bank Policy states "The bank experts, the borrower to take the views of affected groups and local NGOs fully into account in project design and implementation and in particular in the preparation of Environmental Assessment. Consultations with affected people and local NGOs can take several forms and involve different approaches and methods" (Anon 1989, 1993).

2. ROLE OF NGOs

Role of NGOs is multifaceted and can cover many areas (Fig. 1). These may include: environmental awareness programmes, protection of wild life and plantations, afforestation of waste lands, eco-development activities to benefit the rural poor, Joint Forest Management, environment protection and offering training programmes to the locals to equip them with the technical know-how in the implementation of the above activities for their overall socio-economic upliftment. Apart from these, they can also play a major role in the implementation of rehabilitation schemes of the developmental projects by closely working with affected people (MoEF, 1994, 1998). Thus, NGOs build a linkage between the project proponents/developers, the Governmental Agencies and the local people. The all too important role that can be played by NGOs is described in detail below:

A. Participatory Role of NGOs in Environmental Issues Relating to Developmental Projects

With abundant natural resources, a number of developmental

FIG. 1

Role of NGOs in Environmental Conservation

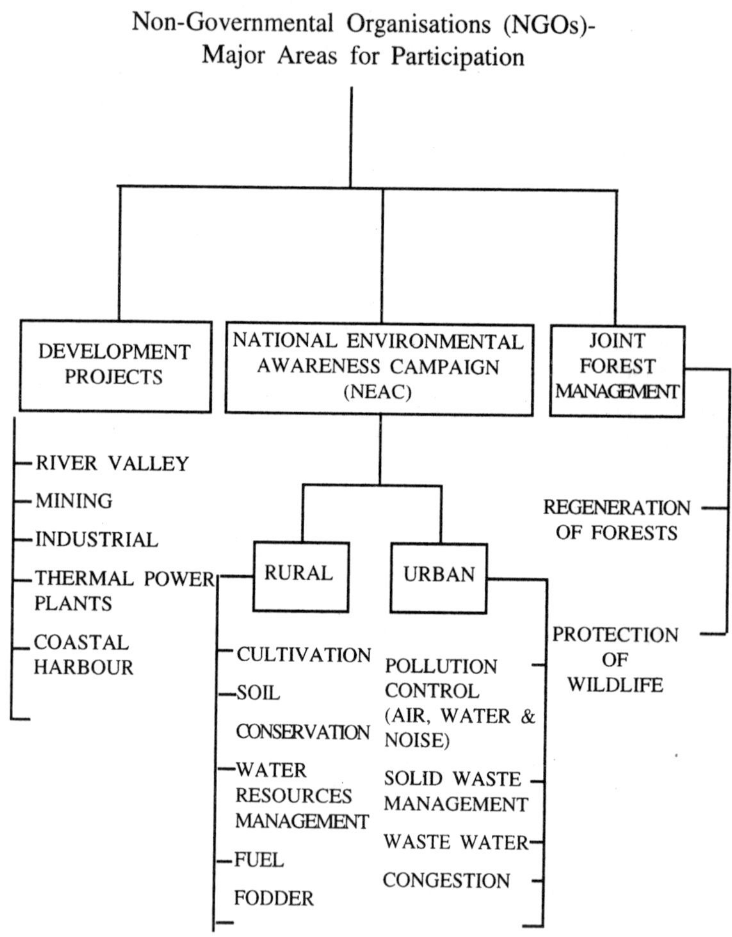

projects of various kinds are coming up in Orissa. As per the report of Eastern Regional Office, Ministry of Environment and Forests, Bhubaneswar (1997), a total of 70 projects have been accorded environmental clearances in the State of Orissa till 1997 by the Ministry of Environment and Forests. These fall into River Valley, (12) Mining, (25) Industry, (13) Thermal, (8) Coastal Harbour, (5) Nuclear (2) and others (5) categories.

In the case of industrial and mining projects, the environmental problems include air pollution, water pollution, noise pollution as well as degradation of soils. The project authorities are expected to take up all mitigative measures and comply with the conditional clearances under Environment (Protection) Act, 1986. The involvement of local NGOs, particularly in the implementation of rehabilitation and checking the various sources of water pollution of streams and rivers will be helpful to serve as a feedback to the project authorities and to take further corrective measures, if needed. Although Orissa does not have severe environmental problems as seen in other industrialised States, the Angul-Talcher area has been identified as one of the 24 problem sites in urban areas in the country by the CPCB for initiation of an action plan for control of environmental pollution (MoEF, 1998). The Central Pollution Control Board has created an NGOs Cell to undertake various programmes to encourage participation of masses in pollution control activities. This Cell provides financial assistance and training to NGOs to carry out Pollution Abatement Programmes in their respective areas. The local NGOs in the State should utilise this facility and equip themselves properly so that they can actively participate in pollution abatement programmes. Besides, in case of all developmental projects, rehabilitation and resettlement of the Project Affected People (PAP) is one of the major problems to be tackled carefully by the project authorities. The involvement of NGOs will greatly help in speedy implementation of the rehabilitation schemes. The role of NGOs in this venture is one of close involvement with the local people in creating proper awareness of both positive and negative impacts of developmental projects and create a harmonious relation between the project authorities and the local people. In fact, Environmental Impact Assessment Notification of 1994 and subsequent amendment in 1997 indicate that representatives of NGOs be included in the Expert Committee for Environmental Impact Assessment.

Through a Government Resolution dated 5.12.1996, the Ganga Action Plan Phase-II has been merged with the National River

Conservation Plan (NRCP) which covers 141 towns located along 22 inter-State rivers in 14 States. Under this, several towns in Orissa are covered along the rivers and water quality monitored to Sambalpur, Cuttack along the Mahanadi, Rourkela and Talcher along the Brahmani and Joda, Jajpur and Chandbali and others along Baitarani. A total of 30 stations have been established and water quality parameters are monitored (CPCB, 1995). The NRCP envisages involvement of reputed NGOs to create environmental awareness through exhibitions, seminars, padyatras, shramdans with special emphasis on the water pollution and remedial measures to be adopted to reduce the impact of water pollution on the health of the citizens (MoEF, 1998).

Orissa is one of the maritime States with over 480 km. coast lines. There are a number of developmental activities along the coast, particularly construction of hotels and tourism resorts along the beaches, expansion of port at Paradeep. Besides, a proposal to set up a port at Dhamra is also under active consideration of the Government. Orissa coast also has been famous for the mass nesting of Olive Ridley turtles (*Lepidochelys olivacea*) at 3 sites. Gahirmatha is considered to be the World's largest sea turtle rookery. Besides, at Devi river mouth and also Rushikulya river mouth mass nesting has been observed (Anon, 1997). With the Coastal Regulation Zone (CRZ) Notification of Ministry of Environment and Forests in 1991 and subsequent amendments, the developmental activities along the coast line invite the provisions of this Notification. The Government of Orissa has also prepared Coastal Zone Management Plan required under the Notification. The NGOs that the active in environmental conservation should pay special attention to the CRZ Notification and create an awareness among the inhabitants along the coast line about the provisions of CRZ Notification and help prevent illegal activities such as mechanised fishing in the coastal waters, unauthorised prawn farming and constructional activities within the CRZ. The NGOs should also be vigilant on these activities as well as the disturbances caused to the natural and sensitive eco-systems by the human interference, particularly at the breeding sites of Olive Ridley tutrles and provide information to the law enforcing agencies in time, so that mitigative measures can be implemented.

B. Involvement of NGOs in Environmental Awareness Campaign

The Ministry of Environment and Forests (MoEF) launched a scheme of National Environment Awareness Campaign (NEAC) in

1986 to create environmental consciousness at all levels of society. NGOs alongwith other organisations, academic institutions, human and youth organisations, Government Departments have been provided with financial assistance for organising seminars, workshops, rallies, training courses, public meetings "Pada Yatras", exhibitions, essay/painting/poster competitions for school children, tree plantation drives, folk dances, street theatres etc. (MoEF, 1994). The MoEF also provides a theme for every year of which the campaign is centered around. During 1991, the major theme of NEAC was "People's Participation in Global Concerns" (MoEF, 1992). Environmental programmes can not be effectively implemented by the Government alone. This is particularly so because of the vastness of the area of the country and also the high population density. There exists a significant variation of environmental issues from urban to rural set up and within the rural areas from one locality to the other in different States. It is, therefore essential for the NGOs to initiate awareness campaigns on the local environmental issues and educate the people of urban and rural areas. The NGOs also have a definite role in imparting education, particularly to the rural folk in dealing with the day to day requirements of drinking water, fuelwood and fodder.

C. Involvement of NGOs in Rural Upliftment

Majority of the villages, particularly those situated in hilly areas are facing problems such as drinking water crisis, scarcity of firewood and fodder. In addition, provision of proper food security, material for construction of houses need to be ensured. Besides, the tribals living in hilly areas practice slash and burn/shifting cultivation, particularly in the districts of Koraput and Malkangiri. The hilly terrain has an inherent problem of soil erosion due to removal of vegetation to meet firewood requirement as well as due to practice of shifting cultivation. Due to this, inspite of good amount of rainfall in these areas, all the water drains off quickly and maintenance of the soil moisture in the slopes becomes difficult and thus this practice has become unsustainable both economically and ecologically. These problems need to be addressed by the NGOs at the village level by closely involving the village communities, particularly the woman. It is necessary to organise awareness programmes highlighting the reasons for these problems in the villages and offer suitable training to the villagers to tackle them. NGOs need to make innovative measures through undertaking pilot projects in the sensitive areas with the financial assistance of different funding agencies from within the country and also abroad. The experience and knowledge gained through implementation of such projects should be disseminated to the villagers by organising workshops and orientation programmes that provide better management practices, cultivation techniques and other related activities to benefit in the over all development of the villages. In order to make the cultivation more productive, NGOs may initiate appropriate measures by involving the locals. The measures may include construction of check dams, stone walls and contour bunds in sensitive locations to help retain soil. By considering the slope factor and also availability of water source, small pondages can be built for storage of water. The NGOs also may encourage settled agriculture in hilly areas in place of shifting cultivation. This, perhaps may require various soil conservation measures such as proper levelling and terracing, contour bunding, construction of stone walls in sensitive zones, gully plugging, vegetative fencing and plantation along the streams. Some field demonstration projects may be worth initiating on a priority basis. NGOs should build a partnership with the locals by involving them closely holistic that ultimately helps in development of the area by adopting suitable practices. The farming techniques may include mixed cultivation and growing fodder species on the farm bunds which will not only serve in soil conservation but can be used as fodder for livestock. The farmers may also be benefited by cultivation of fruit trees on the bunds. Depending upon the extent of the area and its suitability, plantation of local tree species and their protection should be encouraged so that the hills are covered with vegetation. The NGO's role, thus, would be to involve locals in the planning process, local resource utilisation and to develop the area to meet the requirements of foodgrains, drinking water and firewood as well as fodder for the livestock. The NGOs also should promote peripheral activities such as literacy improvement, health education, development of proper communication as well as marketing facilities to sell the produce from the villages. Depending upon the situation, NGOs may also provide technical information in the development of pisciculture, aquaculture, piccary, poultry etc. Through the above activities, NGOs definitely can contribute to the over all development of villages that are the backbone of the country's economic progress.

D. NGO's Role in the Mitigation of Environment Problems in Urban Areas

The environmental problems that confront man in the urban areas include pollution of air and water environment, solid waste management, waste water discharge, industrial pollution and urban congestion are major environmental issues that deserve high priority for India (UNEP, 1997). With the industrialisation, there has been a phenomenal growth of urban centres, consequently leading to congestion. The combustion of fossil fuels in the industrial and transport sectors has been the major source of pollution, posing potential health hazards (Anon, 1998b). Besides, vehicular traffic in urban areas also create noise pollution. Solid wastes generated in cities adds a new dimension which needs to be tackled.

There are a number of laws that have been enacted and institutions created to control and regulate environmental offences. The Constitution of India has also empowered various Bodies to take suitable measures for protecting the environment. As has already been mentioned in the introduction, under the Environment (Protection) Act, 1986 and other laws such as Air (Prevention and Control of Pollution) Act, 1981 and Water (Prevention and Control of Pollution) Act, 1974 and various amendments thereof, provisions have been made for control and abatement of environmental pollution. Besides, the Ministry of Environment and Forests (MoEF), the Appex Administrative Body at the Central Government, Wildlife, National River Conservation Directorates and National Afforestation and Eco-development Board have been set up and given the responsibility to elicit various conservation plans and their implementation in the country. The Central Pollution Control Board and the State Pollution Control Boards are the Administrative Bodies for exercising regulatory functions under the Acts mentioned above.

In the State of Orissa, the pollution is limited only to a few pockets. The Central Pollution Control Board (CPCB) identified Angul-Talcher area as one of the Environmental Hot spots in Orissa. The Orissa Pollution Control Board (OPCB) has reported that around 76,587.6 Kilo litres of water is discharged every day by 26 major towns and urban conglomerates in Orissa (OSPCB, 1998). In the urban centres, the number of vehicles, including two wheelers, has been on the increase, particularly during the last decade. In 1973, there were 58,226 vehicles registered which increased to 1,63,469 in 1983 and 5,18,848 in 1993, an increases of 270%. With this, there is an increase in the gaseous emissions as well as noise levels. Over 3,86,000 of these are 2 wheelers and 3 wheelers. The number of vehicles is still on the increase. The air emissions as well as noise levels pose a potential threat to the health of urban population (OPCB, 1996). The OPCB has set up 7 National Ambient Air Quality Monitoring Stations with the financial assistance from CPCB. These are located at Angul-Talcher area, Rourkela and Rayagada. Besides, OPCB has fixed 4 stations under the project "State Ambient Air Quality Monitoring (SAAQM)" at Bhubaneswar and Cuttack (OPCB, 1999). The parameters such as SPM, SO_2 and NOx are being regularly monitored. Various Local Bodies have been directed by the OPCB to prepare action plans for disposal and management of solid wastes. The OSPCB is also setting up a vermi-composting project at Khurda to handle 500 kg/day of organic wastes. It has also published a booklet Vermi-Biological for the benefit of Local Bodies. Besides, it has also requested various medical colleges, hospitals and nursery homes to form an action plan for management of bio-medical wastes under the Bio-medical wastes (Management and Handling) Rules, 1998 of MoEF. Whether it is the problem of air and water pollution in industrial areas or mining areas, in cities or it is the problem of dealing with the solid wastes generated, the law enforcing authorities alone can not solve the problems without the cooperation of the people. To make the people aware of the health risks due to air and water pollution as well as proper disposal of solid wastes, the OPCB initiated various awareness programmes. These need to be still strengthened further by the active involvement of the NGOs and organising awareness programmes so as to take the message to all sections of society that contribute to the above problems either directly or indirectly. It is also time for the NGOs to have a closure interaction with the Government Agencies and with the prospective entrepreneurs and motivate them to set up plants to treat the organic wastes by adopting suitable technology. It is also important that the NGOs should keep a strict vigil on the violations of various provisions of environmental Acts by agencies connected with developmental activities and bring such violations to the notice of the enforcement authorities to initiate punitive measures.

E. Involvement of NGOs in Joint Forest Management (JFM)/ Participation Forest Management (PFM) and Wildlife Protection

The National Forest Policy Resolution mentions that the forests in the country have suffered serious depletions which is attributable to relentless pressures for fuelwood, fodder and timber, inadequate protection measures, diversion of forest land for non-forest use and the tendency to look upon forests as a source of revenue of Government. Besides, it is a fact that the village communities living closeby the forest areas have not got the desired benefits of the forest management practices under government control. Due to this, the destruction of forests continued as people viewed that the forests belong to the government and it is the responsibility of the government to undertake plantations and protect existing forests. Subsequently, the local people developed some sort of insecurity to meet their demands from the forests on which they are totally dependent for their day-to-day requirements. This led to struggle for control of resources by people. The changing scenario has been duly recognised by the government in time and a new National Forest Policy has been drawn in 1988 which envisages involvement of village communities in the protection of forests and regeneration of degraded forest lands and share benefits accrued and meet the basic requirements of fuelwood, fodder and small timber for house construction by the people living close by the forests (MoEF, 1998 b. Anon, 1993). The Policy also aims at restoration of ecological balance and conservation of flora and fauna and maintenance of biological diversity. To achieve these objectives, a number of measures have been proposed of which soil conservation, treatment of catchment areas of rivers, lakes/reservoirs, conservation of water resources, mitigation of floods and droughts form a few of the measures. Committed NGOs may prove well suited to motivate the people and organise the villagers in this task and bring them close to the government in realising the objectives of JFM.

There are about 30 million ha of degraded lands in the country which are under control of the State Forest Department (Shah, 1994). As per the 1997 FSI Report, there are 26.1 million ha of open forests in the country of which 20,629 ha is in Orissa State (Anon, 1998 b). These once have had a good forest and wildlife. However, due to pressure from people from the adjoining areas as described above, and a number of other factors have contributed to the degradation and denudation of these forest areas. The trends continue to be so as peoples' dependence on forests has been continuously on the increase, particularly for purposes of grazing their cattle, for meeting the firewood requirement of the villagers, for minor forest produce for house construction purposes. Regeneration of forests in such degraded areas by the Government requires heavy expenditure, particularly because it involves protection from further destruction and against future encroachment by the villagers. However, as foreseen by the new National Forest Policy, if the villagers themselves come forward and offer to protect such areas and share the benefits among them, the task becomes much easier. Involvement of villagers in regeneration, protection, management of such areas will help in rehabilitation of degraded forest lands. For achieving success in the regeneration of forests, establishment of a grass-root community/institutions such as the Forest Protection Committees (FPC) to shoulder the responsibilities under Joint Forest Management along with the State Government Departments is necessary. The NGOs can facilitate and promote forest restoration activities by motivating the villagers and bridge the gap between the villagers and State Forest Departments and participants in the FPC. The NGOs should have sufficient technical knowledge and managerial skills in all activities from nursery raising to maintenance of plantations and inculcate interest among the village communities. The role of NGOs, thus, will be not only to work in close contact with the local villagers and with the State Forest Department but also to take initiative in the creation of proper awareness among all the villagers and elicit scientific management plans in the regeneration of degraded forests and in proper utilisation of these areas to meet their domestic requirements on sustained basis.

Involvement of locals in the protection of wildlife harbouring the forest areas, particularly near the Wildlife Sanctuaries and National Parks is also important in the conservation of genetic diversity. This can be encouraged through participation by locals in the eco-development programmes. NGOs should impart proper knowledge of the importance of conservation of genetic diversity to the village communities and encourage them to prevent poaching of wildlife.

3. CONCLUSION

The role of NGOs in various fields has already been recognised world-wide. More and more NGOs are being registered anew with a commitment to involve in various nation building activities including conservation of natural resources and protection of environment, which is an encouraging sign towards achieving a better environmental quality in the country. Although some of the NGOs have initiated a leading role in the protection of environment, there is still a larger scope for further improvement in urban areas as well as in the upliftment of rural population. NGOs in general, have assumed the role of implementing various mitigative measures themselves. This role should be supplemented with the role as a promoter/facilitator, particularly in the rural set up so that the local population itself will take up the implementation of schemes related to environmental protection, nature conservation and management. Such an approach will go a long way and make the locals self-supporting in the absence of NGOs, later on. The NGOs also should provide a linkage and liaison between various local groups and the Government Agencies in the implementation of appropriate schemes and in proper utilisation of local resources base. Their role in making people aware of environmental related problems need not be over emphasised. With clear aims and objectives, the NGOs are expected to play a pivotal role in the overall development of environment to benefit all sections of society and help generate a quality environment and nurture it for posterity.

References

Anon, 1989, *Environmental Assessment Source Book*, Vol. 1. Policies Procedure and Cross-Sectoral Issues. The World Bank, Washington, DC. 227

Anon, 1993, *Community Involvement and the Role of NGOs in Environmental Assessment*, 191-270p. In: Environmental Assessment Source Book, Vol. 1. Policies Procedure and Cross-Sectoral Issues. The World Bank, Washington, DC. 227.

Anon, 1993, Public Involvement in Environmental Assessment. In *Source Book Update*, No. 5, The World Bank, Washington, DC.

Anon, 1993, *Joint Forest Management Update*. Society for Promotion of Wastelands Development, New Delhi. 185 p.

Anon, 1995, India's Environment Action Programme—Issues, Approaches and Initiatives Towards Sustainability. Centre for Environment Education, Ahmedabad, 90p.

Anon, 1997, *Orissa At a Glance*. State Forest Department, Government of Orissa, 64p.

Anon, 1998, A *Tata Energy Data Directory and Year Book*, 1998-99, TERI, New Delhi, 348p.

Anon, 1998b, *The State of the Forest Report*, 1997. Forest Survey of India, Ministry of Environment and Forests, GOI, Dehradun.

CPCB, 1995, Annual Report, 1994-95 of Central Pollution Control Board, 135p.

GOI, 1997, Third Conference on Environment and Forests of Eastern Region, December 5, 1997. Ministry of Environment and Forests, Eastern Regional

Office, Bhubaneswar, 208p.

MoEF, 1994, Environment and Development, Ministry of Environment and Forests, GOI. 44p.

MoEF, 1995, Annual Report, 1994-95, Ministry of Environment and Forests, GOI, ENVIS, 136p.

MoEF, 1998, Annual Report 1997-98, Ministry of Environment and Forests, GOI, ENVIS, 172p.

OPCB, 1996, Vehicular Pollution, Orissa Pollution Control Board, Bhubaneswar, 33p.

OPCB, 1998, Parivesh Samachar 5 (1, 2): 9-11, Orissa Pollution Control Board, Bhubaneswar.

OPCB, 1999, Parivesh Samachar 6 (1, 2): 8-11, Orissa Pollution Control Board, Bhubaneswar.

Shah, S.A., 1994, Silvicultural Management of Our Forests. Wastelands News: 9(2): 8-30.

UNEP, 1997, Asia-Pacific Environment Outlook, Thailand. In: Tata Energy Data Directory Year Book, 1998, TERI, New Delhi, 348p.

WWF, India, 1994, Environmental NGOs in India—A Directory, 1994. World Wildlife Fund for Nature—India, New Delhi, 638p.

13

Sustainable Development and CCD

A. JAGANNATHA RAJU

A BRIEF HISTORY OF CCD

The Centre for Community Development (CCD) originated from the long and varied experience of working with the soura tribal communities living in the Gumma hills of Gajapati district, Orissa.

The Founder of the Centre for Community Development A. Jagannadha Raju prior to its formation worked as the staff of another NGO called KMDS and having the gratifying experience in the field of social service to serve the people at the grass root level, developed the insight into the various ways and circumstances in which tribals were exploited in the area and lastly came to the conclusion that unless and until the tribals were organized on their own gregariously as a strong collective force and able to resist their exploitation, no meaningful development could take place in near future. So he himself came forward altruistically to dedicate himself for the betterment of the tribal community, started a Voluntary Organization independently. At last but not least, from the riddled mud of persistent efforts, Centre for Community Development bloomed up for ever to serve the tribals of Gajapati, in their Sustainable Development. The other members included in Centre for community Development are the like minded versatile persons who were conversant with the context of tribal development in Orissa in a holistic manner.

HRD of the Organization

CCD members are having rich expertise of 20 years and at present operating in 100 villages in 4 Blocks namely Rayagada, Gosani, Kasinagar and Gumma of Gajapati District with a staff support of —75 Part Time-Animators,—25 Full Time—Co-ordinators and Cluster Promoters—90 Volunteers supported by Indo Swiss Project Orissa, Action Aid, IGSSS and CARE.

Aims and Objectives of CCD

(a) To help people irrespective of caste, creed, religion and sex in promoting Education, Health, Environment and Legal awareness, Cultural and Socio-economic standards and making the communities as self reliant as possible.

(b) To co-operate, collaborate and co-ordinate with the Government and international institutions that are intended for the up-lift of weaker sections and for the communal harmony and to promote National integration, supplementing the Government efforts.

(c) To mobilize and organize the poor and generate awareness to demand quality services and impose a community system accountability for the performance of village level Government functionaries.

(d) To demonstrate how village and indigenous resources could be used, how human resources, rural skills and local knowledge, grossly under utilized at present, could be used for their own development.

(e) To disseminate information through workshops, and trainings, and other activities to bring about over all development among poor, sick, needy, illiterate, handicapped and the oppressed masses of youth, aged women and children.

(f) To train a cadre of grass root workers who believe in professionalising voluntarism.

(g) To activate the delivery system and to make it effective at the village level to respond to the felt needs of the poorest of the poor.

(h) To promote youth, women associations and co-operatives at the grass root level to support the village community in fulfilling the basic needs.

(i) To promote afforestation, ecological balance and environmental awareness by promoting NRM and Food Security.

(j) To popularize science and technology demystifying them in simpler form to the rural poor.

(k) To carry out the relief and rehabilitation programme to people in areas effected by natural calamities and other emergencies.

(l) To work together with organizations and groups with similar objectives for united action, concerning various education, Socio-economic, legal, cultural, health, women and environment issues.

Relationship

CCD is the leading organization in the District as well as in the State. CCD is the Convenor of District Level Co-ordination Committee of NGO's, (ii) Vice President of District NGO Forum (iii) Convenor of Kalinga Alternation Action Forum leading 43 NGO in the state, (iv) Convenor of Gajapati Vikas Manch Forum leading 13 NGOs of this district level (Common Action Programme). CCD is also a leading Resource Centre to the District NGO.

Development Philosophy (Ideology)

CCD believes that poverty is the direct consequence of unequal distribution of resources where the unjust structures of exploitation and oppression are institutionalized and enforced resulting in the denial of human and basic rights to the majority of people.

The unjust structures have divided the society as the organized elite minority, consisting of rich, literate, politicians and bureaucrats and the other unorganized mass-poor, tribals and Dalits, the deprived majority sector. The organized minority sector is enjoying the fruits putting the unorganized majority sector under exploitation, oppression resulting in an unjust society. We believe that development takes place when these poor unorganized mass organize themselves as a force with critical and alternative thinking and participate in their own development, liberating from exploitation and oppression, leading to structural transformation and empowerment in their favour.

Vision

CCD aspires for a Just Society that is sustainable, participatory, harmony with nature, free from exploitation, oppression and gender basis with decentralized decision making.

Activities of CCD

1. Education and Awareness

- Non formal education for children
- Adult education for elders
- Value based women and girl child education

2. Community Health

- Health awareness education
- Mother and child care/immunization
- Revolving Medical Kit
- Indigenous health care practices
- Integrated Nutrition on Health Programme in the High Impact Block, Rayagada of Gajapati
- Reproductive Child Health Care and special care and attention towards the tribal women and adolescent girls of Gumma Block
- Nutrition-Kitchen Garden mobilization
- Sanitation and safe drinking water

3. Women Empowerment

- Gender perspectives and gender equity
- Women education and gender sensitization
- Self Help Groups—Women savings and credit
- Income Generation Programmes for women
- Women Exposure, women skills development

4. Poverty Alleviation through Income Generation

- Community Savings and Grain Banking System

- Community Plantation
- Goat/Sheep/Poultry rearing
- Bull-Calf Rearing
- Promotion of Dairy/Fishery

5. *Environment*
    - Awareness camps for the protection of environment
    - Nursery Raising and afforestation
    - Forest Protection and restoration

6. *Animal Husbandry*
    - Care and management of Animal Training
    - Animal Health Care
    - Fodder Development
    - Promotion of Dairy

7. *Sustainable Land Development and Food Security*
    - Land Development Programme
    - Watershed Management
    - Sloppy Agriculture Land Technology (SALT)
    - Agriculture Research and Documentation
    - Tank Renovation

8. *Community Organization*
    - Village level, Cluster level and Area level Meeting.
    - Formation of Sanghas with tribal men, women and youth.
    - Training on Socio-economical, Political and Cultural Development.
    - Liasoining with line Departments.
    - Exposure Visits and creation of pressure groups.

THE ROLE OF CCD IN COMMUNITY DEVELOPMENT

For the development of the soura community of Gajapati the role of CCD is very prominent and an outstanding attempt. To empower the soura tribals of Gajapati for the Socio-economic development, to promote the people's participation, capacity and organization and to facilitate programme implementation for the local cadre development is the goal and zenith of CCD indulging itself into the community development.

Poverty Alleviation Programmes

CCD thinks that development of the people is a process which only startes with the people. But in realistic trend CCD feels that the people of the development area are groaning under the fangs of abject poverty, adversity and gross economical imbalance and remaining under the clutches of the supercilious vested groups who maintained the stereo-typing strata, who social inequality and economical dislocation in the development area. So CCD to encourage and develop the sustainability of the peope and to reduce landlessness, extended its help to the poor to access the means of production and to reduce their dependency the money lenders for consumption loans especially during the lean period i.e., from Sept. to Nov. in each year. In connection with these land development, creation of alternative income generation opportunities was also given the top-notch importance to highlight and to upstart the below poverty line people. It also provides the sheep credit scheme where as 125 beneficiaries were provided with the support and 30 beneficiaries were provided with the support of under the calf credit during the year 1992. This year 210 members were given a loan of Rs. 3,01,854 at an interest rate of 12% p.a. To develop the waste land and to provide the sustainable agriculture, agriculture demonstration was done in 50 acres of the land of the target group. It provided the demonstration on sustainable methods of agriculture, like vermi composting and methods of moisture and nitrogen retention.

Community Health Programmes

The development area of CCD is a Malaria endemic zone. The people are inhabiting at the hill top belts are not getting accessibility to medicine facility and the departmental personnel are also not visiting to the hill top areas due to the inaccessibility. Most often the quacks while visiting these pocket-boroughs swindle the gullible tribes by giving them the improper medicines at exorbitant prices. To encounter the situation CCD has taken the bold step in the development area and to creating an implicit demand for health services. It conducted the health and awareness camps at the development area villages and highlighted the existing bottlenecks relating with the health aspects. To practise safe delivery methods 25 trainings were imparted to identified Traditional Birth Attendants and 5 mothers meets were organized to instill awareness among mothers about safe delivery. Importance was also given to take care of the adolescent girls and pregnant mothers. Several meetings and workshops on health awareness and detection of the T.B. Patients were conducted and it brought to the notice of medical authorities after the proper identification of the patients.

Education

Education is the *summum bonum* for the all round and the aesthetic development of an individual probably to say a child's future prospectives. Thus unprecedented importance was given to highlight education in the development area which is riddled under high illiteracy and gross negligence. Therefore CCD has given importance to develop children of the development area by running non formal centres in each village. Literacy percentage of the development area is 6% where as the female literacy is only 2% which is a matter of *ludicrous fiasco* even after decades of planning and intervention of the Government of India. Most probably in the development area the children remained under bondage of the local money lenders. They do not get any scope and opportunity to get the educational facilities. Parents of the children are also poverty stricken and unable to educate their children. They engage them in rearing of cattle and earning money to supplement their family network and income. Therefore the major focus for the year 1999 was to run the NFE centres in the development area villages to help the community access primary education and eventually get assimilated in the formal schools. During the year about 1020 children were enrolled in the NFE centres of CCD where as 248 children were enrolled in the Government and Ashram schools run by the department of tribal welfare, Government of Orissa.

Natural Resource Management

To generate the income source of the people of the operational area villages it was decided to provide various types of sustainable development units in the shape of Land Development, Micro Watershed Management, Sustainable Agriculture and Sloppy Agriculture Land Technology. Land development and creation of alternative income generation opportunities were given emphasis during the year and CCD has taken land development work under drought rehabilitation programme which resulted in more production contributing to food security which has an excellent impact on tribal communities. Land development work ranked top during the year and different land development activities and modalities of action include in the form of stone and soil bunding, guly plugging to check soil erosion. This year about 69.40 acres of land was developed and the fruit bearing trees and usufructs were implanted to provide sustainable food security.

Watershed Management

To provide the efficacious sustainable food security to the tribes CCD has planned to develop the watershed management programmes in order to provide the sustainable source of livelihood to the tribes. The watershed activity is continuing in 10 villages covering 134 tribal families. Land development work which takes the form of stone and soil bunding, gully plugging to check soil erosion are being done by the community by the active co-operation, co-ordination and collaboration of CCD. The watershed stretches more than 700 acres in 4 Blocks of Gajapati District.

Sustainable Agriculture

For the first time in the tribal belts of its operational area CCD was able to introduce sustainable agriculture net work with a view to provide sustainable food security to the tribes and for the first time the demonstration on sustainable methods like vermi composting and methods retention of moistures and nitrogen were adopted. In the demonstration area comprising of 8.15 acres of land during the Kharif seasons hybrid crops like Maize, Ground Nut, Ragi and Arhar etc. during the Rabi season Sun Flower, Black Gram, Potato, Cauli Flowers, Cabbage, Knolkhol, were grown in 25 villages of Gumma and Gosani Block.

A Case Study-Food Security

The target families of the development area obtain low yields from their agricultural lands due to lack of working capital for farm inputs such as fertilizers and pesticides. Though the rainfall is marginalised, the utilization of surface and ground water is very low. The existing waste land and the arid agricultural land are also under-utilized. The forest-based income through minor forest produce and food supplement for the target families provides limited opportunities where as they are facing a lot of trouble in their day to day life. So CCD has taken the initiation to increase the productivity levels of the target families by means of various sectoral approach and strategies in order to provide food security to the target families in a sustainable manner.

The sectoral strategies like management of land and water resources, to increase the productivity of farms and the promotion of subsidiary source of income for the land less tribals is one of the major steps taken by CCD in the sphere of its food security activities. Various programmes are planned for the development of the food security activities. For the development of agriculture demonstration was given in 20 acres of land, i.e. 18 acres of wet land and 2 arcres of podu land of the development area. The underlying activities like meeting on agriculture development, training and exposure on agriculture and supply of seeds and to create water sources are the major activities to provide food security to the target families. For inundation of water source to the arid and dry lands of the area 3 leg pumps were provided to three clusters for vegetable cultivation. An amount of Rs. 15516 has been provided for the supply of ground nut seeds, paddy and other agricultural amenities to 70 beneficiaries of the development area. The vegetable seeds costing about Rs. 2285 has been provided to 30 beneficiaries.

Sloppy Agriculture Land Technology

5 sities for SALT technology were promoted by CCD where as the hedge rows are cut down and incorporated in the land. Gaps in the hedge rows are filled in by the Trofosia and Flemingia during the Rabi Season. Permanent plants were also raised in alternative strips.

Afforestation and Nursery

To nurture and restore the nature is nature of CCD. Basing on this adage CCD has followed the path of its own in an efficacious manner to mobilize for the restoration of the dilapidated forest and it has taken the daunting steps for afforestation programmes and was able to cover more than 2000 acres since its intervention in the area of operation. Protection of environment and to change the village atmosphere nursery raising is one of the most important steps as initiated by CCD to meet the market demand for fruits as well as the demand for the fruit bearing plants from the community, CCD during the year was able to mobilize Rs. 90,000 from DRDA, Gajapati.

A total of 65000 saplings were raised in 6 G.P.s namely Partada, Namnaguda, Khorsonda, M.S. Pur and Bommika G.P.s 42,354 plants were distributed in the operational area villages which covers 270 acres of existing waste land.

SOME CASE STUDIES

(Regarding the Tribal Leadership in the Local Self Government)

Hear our voices . . . .

Janakamma Sabar of Tala Barlanda village is the ward member in the Panchayat. Janakamma has refused to be relegated to the position of a mere rubber stamp. She has used her power and position for the welfare of her fellow tribal. Janakamma mobilized funds from the block for the construction of a road to Tala Barlanda, for the repair of the school building in the village. She played a major role in mobilizing 10 Indira Awas Yojana homes for her village. She is an enthusiastic participant and an important contributor in panchyat meetings. Janakamma is making a mark for herself because of her provocative speeches asking her tribal brethren to awake to the opportunities in horizon. She is also a member of the PDS network which is responsible for the smooth distribution of ration.

Practice of Savings and Thrift

End of Tulsiamma's woes

Tulasiamma of Arli village had mortgaged her land to the village landlords. She was unable to retrieve her land because of the repeated failure of crops. Circumstances compelled Tulisiamma to migrate to Hyderabad along with her family in 1995. But the urban realities of living in unhygienic conditions in slums forced Tulsiamma and her family to return to their village in 1997. But on her return, to add to her endless woes, she found that the house that they had constructed had been destroyed by the hazards of the weather. She joined the village Self Help Group and availed a loan of Rs. 4,000 to retrieve her land from the landlords of Korsanda village. Now the relesed land is cultivated by Tulisiamma and her family and is mortgaged to the village committee. This year out of her harvest of ten bags of paddy, she repaid the village committee, four bags @ Rs. 350 per bag. She has been finally able to settle down in her little village, peacefully.

Health Awareness and Enlightenment

No more quacks

Laxmamma Soboro of Losiripur village suffered two miscarriages. She was anaemic and on the advice of the village quack, the medicines and treatment she followed took her condition from bad to worse. Laxmamma was afriad to think of having a child. But this time when she conceived again, she decided to consult the trained TBAs of the project and not visit the village quack. She was referred to the District Hospital by the TBAs and was regularly assisted in her ANC check-ups by the Animators and the TBA from Losiripur. Her baby was delivered by the TBA in Losiripur and she has a healthy child today.

Good Health and Food for All

In the village of Arli, the village women sat together in a group meeting to discuss the health problems that plagued them and their children. In course of discussion, D. Kantamma, a member of the group expressed her opinion very candidly. She told all the group members that they would never be healthy if they continued to eat stale food. She reminded all the members that they all cooked more than they could eat and ate the food which was left over for days. This food was infected with flies. Further the women ate the food which was left over by their husbands and their children and not when it was served for. The food as a result lost its nutrition value. All the members in the meeting took the decision that they would eat only hot food. Though it appears simplistic, but it must be reiterated that in the soura economy, this takes great significance as far as distribution of food in the house hold level is concerned or if we just take it as merely the nutrition aspect.

CONCLUSION

We have been overwhelmed with peoples' response, encouragement from Government Department and unstinted support from the Governmental and Non-Governmental agencies, wholehertedly during the year.

During its short period of extent of CCD, we feel legitimately proud of the achievements in sustainable development and also feel that our awareness of areas where we have failed will help us in guiding our future action plan.

In Brief, our whole life is torturous and an up-hill-pernicious path, full of many ups and downs. But we are confident in quest of our path in a bonafide manner.

14

Sustained Afforestation: Key to Sustainable Development

C.P. Oberai

Forests play a vital role in the economy of developing countries. A large segment of India's population depends on forests for energy, housing, fodder and small timber. The demands for forest products and services is increasing with the growth in population and economy whereas the forest cover in the country is deteriorating. The increased demand of forest produce, land hunger by the increasing population and poverty are among the main causes of deterioration in forest cover.

The continuous deterioration of natural forests of the country is due to disproportionate withdrawals of forest produce as compared to its carrying capacity. Demand for timber, pastures, and diversion of forest lands for agriculture has put enormous pressure on forests. The apparent alternative of afforestation on non-forest lands has not yet picked up in many states for various reasons. At the same time, Government efforts to promote tree planting on private and farm lands under agro-forestry programmes also have not been able to make the desired impact as the incentives under such programmes fell below the expectations of the people.

Reprinted with permission from Yojana, 41:8, August 1997.

Is it that the people deriving the benefit from forests are not aware of fast shrinking forest resources and that they will be the ultimate sufferer? The opinion on this issue may be divided. Whereas a large population is aware of the adverse impact of deforestation, they are more interested in the immediate consumptive needs rather than conservation of resources in perpetuity. It has been realised that the survival and quality of forests in situations where forests are a means of sustenance depend on:

1. Intensive afforestation on degraded forest lands, non-forest lands, non-forest lands including farm lands under various planting programmes.
2. Association of communities depending on forests in management and development of forests and judicious harvesting of forest resources.

PRESENT SCENARIO

The recorded forest area of the country is 76.5 million ha (about 23 per cent of total area of the country). However, the tree cover is only about 64 million hectare (19.5 per cent of total area) out of which 38.6 million hectare has good forest cover having crown density of more than 40 per cent, 25 million ha is degraded with crown density in between 10 per cent to 40 per cent. Some forest area is virtually blank, even bereft of any root stock. As per the State of Forest Report, 1995, there is a decrease in the forest cover of 507 sq. km. in the country as compared with that in 1993. Even the density of good forests is decreasing, and the situation is getting worse with further degradation of degraded forests. Obviously, this is due to overuse and unmindful exploitation of the forest resources by the people living in and around forests. Though the general condition of forests has deteriorated, the extent of tree cover in the country has been around 64 million ha since 1987, thanks to intensive efforts of afforestation, protection and control through various legislation. A comparison of forest cover assessed in alternate years since 1987 by the Forest Survey of India is as follows:

India has a growing stock of 4.74 billion M^3 with an annual increment of around 87.62 million M^3. However, as in 10 million ha of good forest, falling in National Parks, Sanctuaries, Biosphere

Sl. No.	*Year of assessment*	*Forest cover (Area in sq. km.)*			*Total*
		Dense	*Degraded*	*Mangroves*	
1.	1987	357,686	276,583	4,046	642,041
2.	1989	378,470	257,049	4,255	640,134
3.	1991	385,610	250,482	4,242	640,694
4.	1993	385,576	250,275	4,256	640,107
5.	1995	385,756	249,311	4,533	639,600

Reserves etc., no commercial extraction is permitted, only 12 million M^3 of timber and 40 million M^3 of fuelwood is being officially extracted. The unrecorded extraction for meeting local needs under existing rights and concessions in forest areas may be many times more. The current demand of timber is around 30 million M^3, out of which 8.3 million M^3 is needed for paper, pulp and panel products and 15.4 million M^3 for saw milling, i.e. housing, packaging, furniture, etc. During 1994-95, nearly US$600 million worth of wood and wood products were imported. The total timber requirement is estimated to grow to a level of 60 million M^3 by the year 2010 (20.5 million M^3 for paper, pulp and panel products and 27 million M^3 for saw milling. The present requirement of fuelwood in the country is around 280 million tonnes, which is likely to rise to 356 million tonnes by 2010 A.D. There is thus a large gap between demand and supply.

NATIONAL FOREST POLICY

The principal aim of the National Forest Policy, 1988 is to ensure environmental stability and maintenance of ecological balance through preservation and rehabilitation of forests while providing for fuelwood, fodder, minor forest produce and small timber needs of the rural and tribal population. The Policy envisages that the national goal should be to have a minimum of one-third of the total land area of the country under forest or tree cover. In the hilly and mountainous regions, the aim should be to maintain at least two-third of the area under such cover so as to prevent erosion and land degradation and to ensure stability of the fragile eco-system. In order to achieve the goal, the National Forest Policy calls for a massive need-based and time-bound programme of afforestation on degraded forests, wastelands, community lands and the lands of individuals including agriculture lands, with particular emphasis on the production of fuelwood and fodder. The policy also provides that the land laws should be modified wherever necessary so as to facilitate and motivate people to undertake tree farming and growing of fodder plants, grasses and legumes on their own land. As far as the involvement of communities in management of forests is concerned, the National Forest Policy, 1988 makes a watershed in Indian Forestry by recognizing the role of community in development and protection of forests. The Policy document provides for creating a massive people's involvement, including women for achieving the objectives of the policy. In pursuance of the policy, the Government of India issued detailed guidelines to all States and Union Territories on June 1, 1990 for enlisting peoples' involvement through Joint Forest Management (JFM). So far 17 States have issued administrative resolutions adopting the concept of JFM on benefit-sharing basis. Local people are organised in village level institutions such as village forest committees (VFC) to take part in management of forests.

THE RESOURCES

Forests in the country are subject to very high pressure due to large scale incidents of rural poverty and high population pressure, both human and cattle. It is estimated that the different types of forest products worth Rs. 40,000 crore are withdrawn annually, if all the monetized and non-monetized withdrawals are computed at market value. However, the corresponding investment is only about Rs. 4000 crore (10 per cent of total estimated withdrawals and less than 1 per cent of total Plan outlay of the country). There is hardly any financial resource available to the VFCs for rehabilitation of degraded forests. A holistic view of investment for forest development does not give much hope of accelerated improvement in rehabilitation of forests. More so, the withdrawals from forests are likely to increase at an exponential rate with the increase in population. Earlier, a sizable portion of funds from poverty alleviation programmes was being allocated for social forestry in rural areas but there is no sectoral allotment now under various poverty alleviation schemes. In such circumstances, the community users have themselves to develop a paradigm of sustainable development through creating stakes for people who are presently alienated and have become indifferent to the future of forests. There is need to evolve a system whereby a sizable portion of proceeds from the harvest of forest produce, even if assumed nationally, gets ploughed back in kind (labour/or cash) for rehabilitation of forests. The allocation to forestry sector under successive Five Year Plans and its percentage to the total plan outlay of the country are as follows:

Sl. No.	*Plan*	*Allocation to forestry sector (Rs. in lakh)*	*Percentage of total outlay*
1.	I	764	0.39
2.	II	2121	0.46
3.	III	4585	0.53
4.	IV	8942	0.54
5.	V	20884	0.51
6.	VI	69249	0.71
7.	VII	185910	1.03
8.	VIII	400000	1.00

In order to mobilize funds for development of forestry sector, as well as to address forestry related issues, a National Forestry Action Programme (NFAP) is being prepared with assistance for UNDP and FAO, which is expected to be completed by September 1997.

NEW INITIATIVES

To achieve the targets set out by the National Forest Policy, 1988, an additional area of about 45.5 million hectares needs to be afforested and 25.0 million hectares of degraded forests would need restocking, presuming that there is no further deforestation. Demand-based assessment of afforestation requirements was assessed in 1985 as 5 million hectares per annum and the National Wasteland Development Board was constituted to achieve this goal. To take afforestation at this level, forestry infrastructures need to be strengthened and expanded. As against the need of afforestation as stipulated above, present afforestation efforts are not only very low and inadequate but also show a declining trend, mainly because of less funds earmarked for afforestation.

During the period from First to Sixth Five Year Plan, 8.20 million hectares of land was brought under forests. During Seventh Plan, afforestation was done on 8.88 million hectares. Year wise targets and achievements for afforestation for the two Annual Plans (1990-91, 1991-92) and VIII Plan are given in the table:

Period	*Afforestation target (in million ha)*	*Achievement (in million ha)*
1990-91	1.80	1.39
1991-92	1.80	1.73
1992-93	1.79	1.68
1993-94	1.84	1.52
1994-95	1.95	1.52
1995-96	1.69	1.52
1996-97	1.69	

It is evident that the current requirements and growing demands cannot be met from the present incremental growth and level of plantation. The plantations, mostly government funded, have only reached a level of one million ha per annum in degraded forests and 0.4 to 0.5 million ha in non-forest lands are private lands. This will not meet even the fuelwood needs of the country on sustained basis while the growing demand of industrial wood will degrade the remaining natural forests. Moreover, in view of highly polluting nature of producing wood substitute like aluminium, plastic or steel products as well as shortages of recycled paper, the demand for industrial wood will keep increasing.

One of the primary obstalces in raising of high quality plantations is lack of adequate financial resources. The government funds should naturally go for raising fuelwood, small timber and fooder plantations for meeting the requirements.of local people and tribals who are dependent on forests for their basic needs. In view of the growing gap between demand and available sustainable yield, the government agencies will have to raise around 3 million ha per annum of fuelwood, fooder and timber plantations including the regeneration of felled forest areas to meet the highly subsidized basic survival needs of the rural poor and forest dwellers. The other important obstacle in maintenance of plantations is the increasing interference of human and cattle living in and around the forests.

JOINT FOREST MANAGEMENT

The policy of involving people in management of forests has been adopted in degraded areas. These degraded forests are being developed by associating the local communities in the management and protection of forests as per the policy framework of Joint Forest Management (JFM).

The policy guidelines provide that the bonafide domestic requirements of forest dependent people as regards fuelwood, fodder, non-timber forest products and construction timber should be the first charge on the forest product. The policy document also enjoins that the forest communities should be strongly motivated to identify themselves with the development and protection of forests from which they derive various benefits. However, the problem lies in developing ways and means to identify, and use, the capacity of the community to bring about improvements in the condition of forests. Also the situation and socio-economic conditions in various parts of the country are so diverse that it is impossible to have a uniform policy for community participation. Therefore, the Government of India in its guidelines of June 1, 1990 left it to the initiative of the state governments to devise appropriate strategies and modalities for implementation of community participation, depending upon site-specific situations prevailing therein. People's participation in forest management, euphemistically titled Joint Forest Management (JFM), has since been institutionalised by most of the states. The emphasis has been on formation of village forest committees (VFCs) for participatory management of degraded forests on usufruct sharing basis sans ownership of forest lands. The community would be given free access to usufructs like grasses, lops and tops of branches, and minor forest produce. They would also be given a portion of the proceeds from the sale of final harvest, if they successfully protect the forest through the entire rotation of the tree crops. This varies from 25 per cent to 100 per cent amongst the different states. So far 17 States have issued notification/resolution for the formation of village forest committees involved in JFM. It is estimated that about 2 million hectares of degraded forests are already being managed under Joint Forest Management by about 15000 village forest committees.

Coordinated Efforts

Sustainable Forest Management (SFM) is a wide ranging, comprehensive expression which embraces all aspects of forest conservation and development. Natural resources have to be conserved and augmented but essential development needs of the society cannot be overlooked. Forest management on sustained basis is a herculean task which can well be achieved by joint efforts of the State and people. JFM is a bold step in this direction. Several wood-based industries have come forward to help and participate in overall efforts to accelerate afforestation. NGOs are also actively helping in this task of greening India. Coordinated efforts of all well meaning agencies should result in restoring the green glory that has always been the hall mark of India's forests.

15

Environment Friendly Agriculture

M.S. SWAMINATHAN

The term green revolution coined by Dr. William Gaud of the United States of America in 1968 has come to be associated with not only higher production through enhanced productivity, but also with several negative ecological and social consequences. There is also frequent reference to the fatigue of the green revolution, due to stagnation in yield levels and due to a large requirement of nutrients to produce the same yield as in early seventies.

Is it likely that as we enter a new millennium, we will not have the benefit of new technologies which can help our farmers to produce more food and other agricultural commodities from less land and water?

I believe we are now in a position to launch an ever-green revolution which can help increase yield, income and livelihoods per units of land and water, if we bring about a paradigm shift in our agricultural research and development strategies. The green revolution was triggered by the genetic manipulation of yield in crops like rice, wheat and maize. The ever-green revolution will be triggered by

This article first appeared in *Yojana*, Vol. 41: No. 8, August 1997. Reprinted with permission.

farming systems which can help produce more from the available land, water and labour resources without either ecological or social harm. Thus, progress can be achieved if we shift our mind set from a commodity-centred approach to an entire cropping or farming systems approach. This does not mean that we should decelerate our efforts in the area of crop improvement research. But such research should be tailored to enhancing the performance and productivity of an entire production system. The transition from the fatigue of the green revolution to an ever-green revolution involves a shift from a crop-centred approach to a systems-based approach to technology development and dissemination.

Let us take for example the prospects for "super-rice", capable of yielding over 10 tonnes of rice per ha. Such a rice plant will need a minimum of 200 kg N per ha, together with other major and micro-nutrients. Addition of such nutrients solely through mineral fertilizers will lead to serious environmental problems, and hence, the introduction of legumes in the rotation becomes important.

Scientists now have unique opportunities for designing farming systems for achieving the triple goals of "more food, more income and more livelihoods" per ha of land by harnessing the tools of ecotechnologies resulting from a blend of traditional knowledge with frontier technologies such as biotechnology, informatics including GIS mapping, space technology, renewable energy technologies (solar, wind, biomass and biogas) and management and marketing technologies.

Industrial countries are responsible for much of the global environmental problems such as potential changes in temperature, precipitation, sea level and incidence of ultraviolet-B radiation. While further agricultural intensification in industrialised countries will be ecologically disastrous, the failure to achieve agricultural intensification and diversification in developing countries where farming provides most of the jobs will be socially diastrous. This is because, agriculture including crop and animal husbandry, forestry and agro-forestry, fisheries and agro-industries provides livelihood to over 70 percent of our population. The smaller the farm, the greater is the need for higher marketable surplus for increasing income. Eleven million new livelihoods will have to be created every year in India and these have to come largely from the farm and rural industries sectors. Importing food and other agricultural commodities will hence have the same impact as importing unemployment. Thus, what we need now is an environmentally sustainable and socially equitable green revolution or what may be termed an ever-green revolution.

MEETING THE CHALLENGES

The responses being developed and field tested by the M.S. Swaminathan Research Foundation (MSSRF) to identify implementable approaches at the micro and policy levels to meet the challenges outlined earlier are briefly described below:

(a) Linking the ecological security of an area with the livelihood security of the local community: creating an economic stake in conservation

The community biodiversity programme of MSSRF illustrates how such mutually beneficial linkages can be fostered in biodiversity rich areas. It is a sad fact that the tribal and rural families who have conserved and enhanced biodiversity remain poor, while those who are utilising the products of their efforts become rich. When the conservers have no social or economic stake in conservation, denudation of natural ecosystems becomes more rapid. MSSRF has adopted a three-pronged strategy for creating an economic stake in biodiversity conservation.

First, a transparent and implementable methodology has been developed for incorporating in *sui generis* systems of plant variety protection procedures for recognising and rewarding informal innovations in genetic resources conservation and enhancement.

Second, a symbiotic social contract between commercial companies and tribal and rural families is being fostered for the purpose of promoting the cultivation by local communities of genetic material of interest to the companies on the basis of buy-back arrangements. Such a linkage will prevent the primary material being unsustainably exploited.

Third, local women and men are trained in the compilation of biodiversity inventories and in bio-monitoring, so that they themselves become custodians of their intellectual property. Such trained women and men constitute an Agrobiodiversity Conservation Corps and will be able to help their respective communities to deal with issues such as "prior informed consent" in the use of genetic resources.

Technical Resource Centre

For assisting the community biodiversity movement, MSSRF has established a Technical Resource Centre for the implementation of the equity provisions of the Convention on Biological Diversity. Since this is the first Technical Resource Centre of its kind in the world, the six major components of the Centre are described below:

(i) *Chronicling the contributions of tribal and rural families to the conservation and enhancement of agrobiodiversity* through primary data collection in the states of Tamil Nadu, Kerala, Andhra Pradesh and Orissa as well as in the Lakshadweep and Great Nicobar group of islands.

(ii) *Organisation of an Agrobiodiversity Conservation Corps* of young tribal and rural women and men, who have a social stake in living in their respective villages and who, with appropriate training, can undertake tasks such as compilation of local biodiversity inventories, revitalisation of the *in situ* genetic conservation traditions of their respective communities, monitoring of eco-system health with the help of appropriate bio-indicators and restoration of degraded sacred groves. The members of the corps will be able to assist their respective communities in dealing with "the prior informed consent" provision of the Convention on Biological Diversity in the use of genetic resources.

(iii) *Development of multimedia databases documenting* the contributions of tribal and rural families in the conservation and improvement of agrobiodiversity, for the purpose of enabling them to secure their entitlements from National and Global Community Gene Funds.

(iv) *Maintenance of a community Gene Bank and Herbarium:* A Community Gene Bank with facilities for medium term storage has been established to conserve farmer preserved and developed seeds from the tribal areas of South India. The material will be catalogued and linked to the Technical Resource Centre database. The herbarium serves as a reference centre for the identification of landraces, traditional cultivars and medicinal plants conserved by tribal and rural families.

(v) *Revitalisation of genetic conservation traditions of tribal and rural families* through social recognition of their contributions and the creation of an economic stake in conservation. For this purpose, replicable models of private sector engagement in contract cultivation by tribal and rural families of plants of commercial value are being developed.

(vi) *Legal Advice Cell:* This cell will make available to tribal and rural families appropriate legal advice in matters relating to intellectual property rights and plant variety protection.

(b) The Population Supporting Capacity of Ecosystems: Local Level Socio-demographic Charter

In order to help internalise an understanding of the vital need to restrict population growth within the supporting capacity of land, water, forests and the other components of the ecosystem, training modules have been developed to enable the women and men members of village level democratic institutions to prepare socio-demographic charters for their respective villages. These are local level planning tools designed to assist in priority setting in the matter of meeting unmet minimum needs. A gender code is an important component of the charter. Such socio-demographic charters will help local communities to view population issues in the context of social development and to ensure that children are born for happiness and not just for existence.

(c) Information and Skill Empowerment

For this purpose, the concept of Information Villages has been developed. Trained rural women and men will operate Information Shops where generic information on the meteorological, management and marketing factors relevant to rural livelihoods will be converted into location-specific information. Trained farm women and men themselves become trainers. The computerised extension system adopted in the information shops also help sensitise local families on their entitlements from government and other programmes. Information technologies provide considerable opportunities for value-added jobs in rural areas. While new technologies are important, folk media are often even more effective in reaching the unreached. Hence, folk plays and folk arts and theatre are fully mobilized for achieving information empowerment. For ensuring the success of information empowerment programmes, the information disseminated should be demand-driven and should be locale-specific.

(d) Environmentally Sound Agricultural Intensification, Diversification and Value Addition

This is achieved through participatory research with farm families. Ecotechnologies like integrated pest management and integrated nutrient supply are used. Ecotechnology development involves the blending of the best in frontier technologies with traditional wisdom and practices. Modern science and the ecological prudence of the past can thus be combined.

Ecotechnologies are also practised in aquaculture. Integrated agriculture and aquaculture techniques enhance both farm income and the nutrition security of the household. Whole villages are being enabled to adopt such integrated, intensive farming systems (IIFS). This approach is essential for meeting the triple goals of more food, income and jobs from the available land and water resources. The seven basic principles guiding the IIFS movement are described below:

(i) Soil Health Care

This is fundamental to sustainable intensification. IIFS fosters the inclusion of stem nodulating legumes like Sesbania rostrata, incorporation of Azolla, blue green algae and other sources of symbiotic and non-symbiotic nitrogen fixation and promotion of cereal-legume rotation in the farming system. In addition, vermiculture composting and organic recycling constitute essential components of IIFS. IIFS farmers are trained to maintain a Soil Health Care to monitor the impact of farming systems on the physical, chemical and microbiological components of soil fertility.

(ii) Water Harvesting and Management

IIFS farm families include in their agronomic practices measures to harvest and conserve rain water, so that it can be used in a conjuctive manner with other sources of water. Where water is the major constraint, technologies which can help to optimise income and jobs from every litre of water are chosen and adopted. Maximum emphasis is placed on on-farm water use efficiency and on the use of tehcniques such as drip irrigation, which help to optimise the benefits from the available water.

(iii) Crop and Pest Management

Integrated Nutrient Supply (INS) and Integrated Pest Management (IPM) systems form important components of IIFS. The precise composition of the INS and IPM systems will depend on the components of a farming system as well as on the agro-ecological and soil conditions of the area. Computer-aided extension systems will provide farm families with timely and precise information on all aspects of land, water, pest and post-harvest management.

(iv) Energy Management

Energy is an important and essential input. Besides the energy efficient systems of land, water and pest management described earlier, every effort will be made to harness biogas, biomass, solar and wind energies to the maximum extent possible. Solar and wind energy will be used in hybrid combinations with biogas for farm activities like pumping water and drying grains and other agricultural produce.

(v) Post-harvest Management

IIFS farmers will not only adopt the best available threshing, storage and processing measures, but will also try to produce value-added products from every part of the plant or animal. Post-harvest technology assumes particular importance in the case of perishable commodities like fruits, vegetable, milk, meat, egg, fish and other animal products, and processed food. A mismatch between production and post-harvest technologies affects adversely both producers and consumers. Growing urbanisation leads to a diversification of food habits. Therefore there will be increasing demand for animal products like milk, cheese, eggs and processed food. Agro-processing industries can be promoted on the basis of an assessment of consumer demand. Such food processing industries should be promoted in villages in order to increase employment opportunities for rural youth. In addition, they can help to mitigate micronutrient deficiencies in the diet.

Investment in sanitary and phytosanitary measures is important for providing quality food both for domestic consumers and for export. To assist the spread of IIFS, Governments should make a major investment in storage, roads, transporation and on sanitary and phytosanitary measures.

(vi) Choice of the Crop and Animal Components of Farming Systems

In IIFS, it is important to give very careful consideration to the composition of the farming system. Soil conditions, water availability, agro-climatic features, home needs and above all, marketing opportunities will have to determine the choice of crops, farm animals and aquaculture systems. Small and large ruminants will have a particular advantage among farm animals since they can live largely on crop biomass. Backyard poultry farming can help to provide supplementary income and nutrition.

(vii) Information, Skill, Organisation, Management and Marketing Empowerment

IIFS is based on the principle of precision farming. Hence, for its success, IIFS systems needs a meaningful and effective information and skill empowerment system. Decentralised production systems will have to be supported by a few key centralised services, such as the supply to credit, seeds, biopesticides, and animal disease diagnostics. Ideally, an Information Shop will have to be set up by trained local youth in order to give farm families timely information on their entitlements as well as on meteorological, management and marketing factors. Organisation and management are key elements and depending on the area and farming system, steps will have to be taken to provide to small producers the advantages of scale in processing and marketing.

IIFS is best developed through participatory research between scientists and farm families. This will help to ensure economic viability, environmental sustainability and social and gender equity in IIFS villages. The starting point is to learn from families who have already developed successful IIFS procedures.

It should be emphasised that IIFS will succeed only if it is a human-centered rather than a mere technology-driven programme. The essence of IIFS is the symbiotic partnership between farming families and their natural resource endowments of land, water, forests, flora, funa and sunlight. Without appropriate public policy support in areas like land reform, security of tenure, credit supply, rural infrastructure, input and output pricing and marketing, small farm families will find it difficult to adopt IIFS.

(e) Increasing Farm and Non-farm Employment

The biovillage programme addresses three key areas—preventing resource degradation, improvement of crop and animal productivity and alleviation of poverty. The biovillage programme in progress in villages in the Pondicherry area of India places equal emphasis on off-farm livelihood opportunities and on-farm jobs. This programme avoids a patronage approach to poverty alleviation.

It regards the poor as producers and innovators and helps to build their assets through value addition to time and labour. The basic approach is on asset building and sustainable human development leading to the growth of entrepreneurship.

HUMAN-CENTRED

The programmes are designed on a pro-nature, pro-poor and pro-women foundation. By placing emphasis on the strengthening of the livelihood security of the poor, the biovillage model of sustainable development revolves around the welfare of the economically and socially underprivileged.

It is thus a human-centered pattern of development. The enterprises chosen are based on marketing opportunities. The technological and skill empowerment of the poor is the major approach. Because of the market-driven nature of the enterprises, the economic viability of the biovillage approach is assured. Production and post-harvest technologies and farm and non-farm occupations are brought together in a manner that both producers and consumers benefit.

Biovillages around biosphere reserves would help in providing alternative sources of meeting the day-to-day needs for food, fuel, fodder and other commodities of the families living near such biodiversity rich areas. Also, biovillages near urban areas help to link the rural producer and the urban consumer in a mutually beneficial partnership.

By producing the processed and semi-processed food products needed in urban areas in the villages around towns and cities, the need for the rural poor to migrate to urban centres for livelihood opportunities is minimised. Also, food processing can be used as a method of providing the needed micronutrients by including millets and grain legumes in the food.

Environment friendly agriculture is the pathway to sustainable food and livelihood security. It is based on the non-exploitative use of natural resources. It is based on Gandhiji's advice, "Nature provides for everyone's needs but not everyone's greed". It is the spread of a greed revolution that causes harm to our life support systems. An ever-green revolution based on harnessing solar energy through green plants and adopting environment friendly agricultural practices can alone ensure opportunities for a productive and healthy life for all.

16

Development and Environment

R.K. Pachauri

Fifty years of India's independence have provided the citizens of this country with a great deal of pride in what has been achieved in the past and offer even greater hope for the future. Undoubtedly, critics would find many flaws in the Indian experience, pointing out, perhaps with some validity, that other nations which gained independence and political freedom around the same time have done substantially better in economic terms than India. However, other country situations are perhaps not valid in an assessment of India, because not only are demographic and natural resource issues of vital importance in determining the pace and structure of economic development, but external political factors also have a major influence. It could be argued, for instance, that if India did not have to go through the trauma of partition (or once having been through that experience if there was a spirit of accommodation and cooperation between India and its neighbours) the resources that have gone into defence preparedness and in countering aggression on several occasions, could have been diverted for the purpose of development. Yet, the record of economic growth in India, particularly in the 1980s and 1990s has been generally satisfactory, even though there have been

Reprinted with permission from Yojana, 41:8 August, 1997.

a few years that saw a sharp decline in growth rates particularly in the period 1989-91. In the current year too, concerns are being voiced on the sustainability of the growth rates achieved since 1992-93. There is a significant decline in the rate of growth of industry and major bottlenecks appearing in the infrastructure sectors.

Conventional measures of growth take into acount only the market value of goods and services produced. However, there are several transactions and changes in assets that take place through activities that are not observed in the market, even though they may substantially affect a country's production function by shifting it in one direction or the other. The stock of natural resources in any society and the state of their health is a vital determinant of productivity in several sectors of the economy. The record of 50 years as an independent nation generally indicates that India has not done very well in maintaining its natural resource base. It is for this reason that TERI launched on Earth Day 1995, a major project called Growth with Resource Enhancement of Environment and Nature (GREEN-India 2047). The purpose of this project is to estimate the damage to and degradation in India's key natural resources and to assign economic values to the change in their quality and quantity. The overall intention of the exercise, however is to develop and articulate alternative strategies for growth and development for the next 50 years of India's independence, such that not only can the damage of the past be fully reversed but that the country is actually able to enhance its environmental and natural resources over this period. Needless to say, the methodology by which physical changes in natural resources can be converted into economic estimates is as yet an area of great complexity, and therefore, any economic value that is arrived at is a function of the generally wide range of assumptions lying behind it. Consequently, an exercise of this nature cannot be conclusive and final. Refinements and improvements in any set of estimates produced would have to take place on a regular basis, as better information and greater knowledge on the issues involved is established.

COMPLEX JOB

The complexity in arriving at economic estimates related to environmental damage is essentially on account of the need to use indirect methods of estimation. Environmental costs generally fall into two categories. First, public health impacts are caused by air and water pollution etc. Secondly, there are productivity changes on account, for instance, of increases in water supply costs, soil degradation, deforestation including reduced density of tree cover. A recent World Bank study by Carter Brandon et. al. has attempted to estimate the economic value of the damage that India is incurring on account of environmental pollution. This was estimated at a figure of 4.5 per cent approximately of the country's GDP. The average in most other nations is between 2 to 2.5 per cent. TERI's own estimates turn out to be much higher because some relevant factors were not included in the former study. For instance, the Carter Brandon study has underestimated the impacts of indoor air pollution. TERI has carried out extensive research on indoor air pollution including assessment of emissions from different types of fuels and cookstoves, and measurement of exposure levels for people living in different types of dwellings. As a result of these emissions measured and the dosage recived by the humans thus exposed, measurement of health effects was also carried out in collaboration with epidemiologists. TERI's estimate is that approximately 2.2 million persons lose their lives annually resulting from high levels of indoor air pollution. The most vulnerable sections of our population in this context are women and children. This indicates that indoor air pollution is far more harmful in India and, therefore, far more serious a problem than outdoor air pollution. This fact also highlights the importance of an appropriate cooking energy policy for the poorest people in rural as well as urban areas.

Even higher than the cost of air pollution is the cost attributable to pollution of water in this country both on the surface and below the ground. The overwhelming majority of diseases prevalent in India arise out of contaminated and increasingly polluted water. Diseases such as diarrhoea, trachoma and hepatitis are the result of extremely high levels of water pollution. The generally high levels of infant mortality are also related to the lack of clean water, because infants reeive a whole range of infections and parasites into their bodies at highly vulnerable ages, transmitted through polluted water. As yet, our monitoring and policing of water pollution is not adequate, and even where measurement takes place regularly, the facts thus obtained are not adequately highlighted. The institutional arrangements for enforcement at the grassroots level are as yet less than effective.

CORRECTION NEEDED

At a general level, the prolonged depletion and degradation of our natural resources stems from a universal preoccupation with measures of Gross National Product (GNP) and Gross Domestic Product (GDP). A correction of this trend can only come from internalizing externalities which do not show up in national income accounts. Wealth is not merely a function of assets produced in a market economy, but consist overwhelmingly of natural assets such as forests, clean air, healthy soil and clean water. Unless an inventory and assessment of these are carried out on a continuous basis, measures of economic output and wealth would remain incomplete and erroneous.

It is not as though economists have always ignored the question of preserving the environment and maintaining a healthy natural resource base in an economic system. Over six decades ago Harold Hotelling first analyzed, within an economic framework, the question of depletion of exhaustible resources. He came up with an analytical framework within which prices and economic values should be arrived at for optimal decision concerning the exploitation of exhaustible resources. Hotelling's work dealt essentially with the depletion of mineral resources, but the framework he established would apply quite aptly to clean air, water and soil as also forests. After Hotelling's time, two pioneering economists Kenneth Boulding and Nicholas Georgescu-Roegen provided a stronger analytical base to the issue of environment and development. Boulding focussed on the production of goods and services, and how these often lead to the production of "bads". Georgescu-Roegen used an analogy from physics to develop the hypothesis that the economic process produces entropy in the system. Experience in several countries indicates an increase in entropy in the economy upto a point after which generally with higher income levels and proper development of institutional and legislative measures, a reversal in this trend is achieved. But this evidence of a turning point is very recent, generally over the past 25 years or so. As forests vanish and the stock of natural resources per capita is depleted such as in the form of reduction of rich soil nutrients, falling of the water table and the growing scarcity of clean air and clean water, the costs associated with economic activity increase sharply.

In recent years research has also been carried out on the question concerning how and when a society reverses the trend of damage that industrialization and the conventional form of development brings about. This question is often answered by investigating what is now known popularly as the environmental Kuznets curve. The typical shape of this curve and differences in its positioning are shown in Figure 1. Essentially a lower position of this curve can be achieved in a society which has well defined property rights, the use of market based instruments and decision making systems whereby environmental costs are internalized, and wherein, last but not least, cultural factors and tradition lead to a lower intensity of use of natural resources for producing specific levels of output of goods and services. Indeed, for a country like India, a great deal of emphasis needs to be placed on how the environmental Kuznets curve could be brought as low as possible, so that the country does not have to wait for reaching high levels of income in order to bring about an improvement in environmental quality. In fact, there may be reason to believe that in the case of India the initial portion of the environmental Kuznets curve is relatively flat, simply because the carrying capacity of the land has been stretched to a point where the extreme levels of poverty that exist and the high level of dependence on basic natural resources that the poor have for their sustenance produce environmental damage of high magnitude.

VITAL OPPORTUNITY

Science and technology provide a vital opportunity for reducing environmental damage in the pursuit of development and economic growth. Conventional technologies employed in the process of modernization of a country like India have come from societies which have not valued the environment appropriately and, therefore, have exploited the environment far beyond optimal levels. While, several of these societies have in recent years corrected this situation by diverting significant resources for both end-of-pipe solutions as well as pollution prevention, a country like India would find it difficult and expensive to change existing capital stock and technologies with this objective. It is, therefore, far more prudent for development policy in our country to favour technologies that prevent pollution in general. This, however, will not take place unless we remove protection of domestic industry and unless we internalize the cost of environmental damage in the production and consumption of goods and services. Of course, the achievement of a system that transparently includes environmental costs and prices is a complex challenge. The use of

FIG. 1

The Environmental Kuznets Curve : The Relationship between Policies, Prosperity, and Environmental Damage

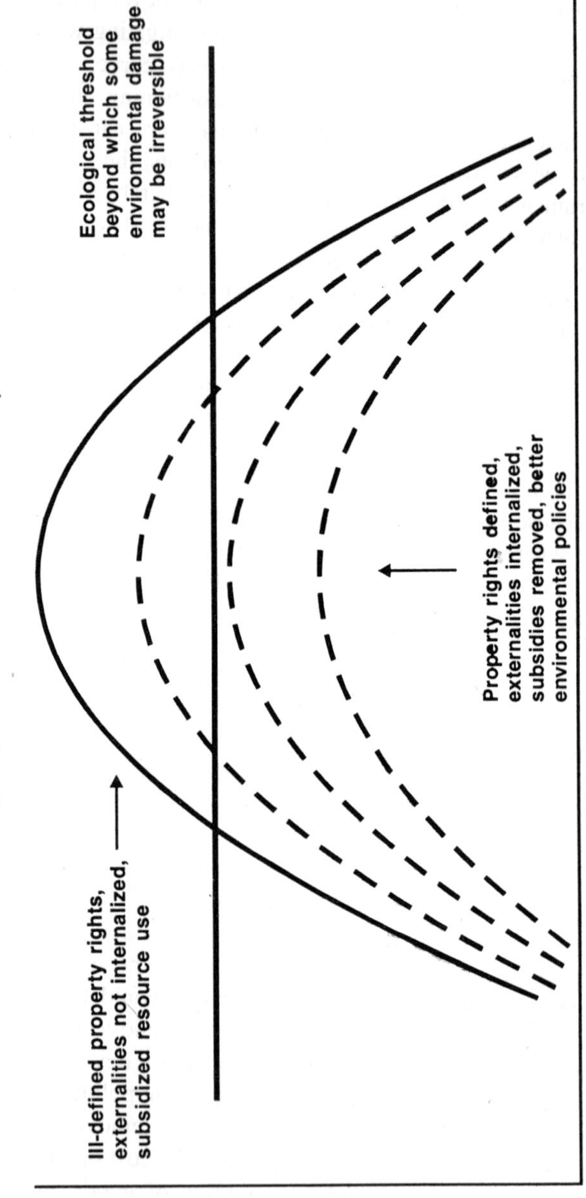

Source : Panayotou (1995).

proper economic instruments to prevent pollution presupposes the existence of measurement and monitoring to effectively ensure that different levels of pollution are accounted for in appropriate levels of pricing. Such a system also presupposes that legal and contractual costs would be low enough in relative terms to ensure proper compliance with legal and regulatory requirements.

EFFECTIVE REGULATION

Effective regulation is an important forerunner of proper technology development. The development of fuel efficient automobiles round the world, particularly in the wake of the first and second oil price shocks, has come largely as a result of strong regulation and mandatory requirements established by governments, as in the case of the United States in the 1970s. The use of unleaded petrol and fitting of catalytic convertors in automobiles has also been driven by regulation, resulting in much cleaner air in countries and cities where this has been the case. However, regulatory measures must be based on solid economic rationale, which essentially requires an assessment of costs and benefits from compliance versus non-compliance. Subsequently, regulatory standards require regular and continuous review, because influencing conditions change every time and in a manner that may not be anticipated at the stage when the standards were set initially.

It is becoming increasingly clear that there is really no conflict between development and protection of the environment. In fact, destruction of the environment increases costs and imposes constraints on development that have been ignored for quite some time in this country. It is now necessary to embark on a rational strategy by which the apparent conflict between the two is resolved. This would, no doubt, require major restructuring of institutions dealing with pollution prevention and control particularly at the grassroots level and an integration of environmental concerns with economic policy. A major effort to educate the public would also be essential to sensitize decision makers and consumers at large on the economic impacts of environmental damage. But a great deal of research and assembly of data must precede the labelling in economic terms of production and consumption decisions in an economic system. It is hoped that the GREEN India—2047 project would be a major step in taking knowledge on these subjects in the right direction.

17

*Tribal Agriculture, Employment and Productivity of Labour in Different Ecological Settings**

JAGABANDHU SAMAL

The tribal economy in India has been studied mainly by anthropolists and administrators; few economists have gone into it in detail. A study of the available literature suggests that there are considerable variations among tribal economies depending upon varying ecological settings. In every ecological setting the tribals pursue more than one line of economic activity. But little factual information is available in these studies about the pattern of employment among tribals, in the sense of how the tribals engage themselves in these multiple economic pursuits at various levels of efficiency. While general statements have been made about the tribals maxmising their returns under given conditions, no empirical data are available to substantiate the proposition. No clear picture emerges about the relative importance

---

* This paper is based on the author's draft Ph.D. Thesis, prepared at the Gokhale Institute of Politics and Economics, Pune, under the supervision of Prof. (Dr.) Neelakanth Rath and Submitted to the University of Poona for award of Doctoral Degree. The title of the Thesis is "Some Aspects of Tribal Economy—A case Study of Koraput District (1992)".

of different activities in different ecological settings.

An attempt is made here, therefore, to go into such unexamined hypotheses and problems by trying to understand the nature of tribal village economy under different ecological settings. It tries to study the pattern of employment of tribals, measure the relative importance and efficiency of such operations, the economic rationale of such patterns of employment and also the seasonality of employment in order to see its relation with economic activities and their efficiency, so that any alternative proposals for improvement should take these into account.

The tribals in India, who constitute 7.76 percent (1981) of total population, with nearly 94 percent of them living in rural areas, are found to live broadly in three geographical divisions. These are (i) the Himalayan region, including the north eastern Zone, in the mountain valleys and the eastern frontier; (ii) the central belt of older hills and plateau along the dividing line between peninsular India and the Indo-Gangetic plains and (iii) the extreme corners of South-Western India and the corverging lines of the Western and Eastern Ghats. These, of course exclude the Andaman and Nicobar, Laccadive, Minicoy and Aminidive Islands. Out of these, it is in the central belt of the country that the maximum concentration (67 percent) of the tribal population is found and, therefore, this region may be called the "Tribal Heart-land" of India. On the basis of ecological differences we may divide the central tribal belt of the country into three separate sub-divisions. These are; (a) the Eastern-central tribal belt, which starts from the northern part of East Godavari district and covers the north eastern coastal districts of Andhra Pradesh, the tribal districts of Orissa, the Western districts of West Bengal, South Bihar, eastern part of Madhya Pradesh boarding Jabalpur, Mandala, Belaghat upto Bastar and Chandrapur, Gadchiroli and Bhandara districts of Maharastrata. This area contains about 40 percent of the total tribal population of the country, and 59 percent of the tribals of the central belt. (b) The second sub-division is the Vindhya-satpura Range along the Narmada and Tapti Valleys, which covers the tribal concentrated districts boardering Madhya Pradesh and Maharastra, starting from Seoni and Chhindwara of Madhya Pradesh and Amaravati of Maharastra and Westward till Thane districts of Maharastra and Valsad, Surat, Bharuch and Vadodara districts of Gujarat. About 16 percent of the total tribal population of India lives in this region. (c) The third one is the North-

Western Dry Region along the Aravali range which includes some of the bordering and contiguous districts of Madhya Pradesh, Gujarat and Rajasthan. These districts are Jhabua, Ratlam, Panchamahal, Chittaugrah, Udaipur, Sabarkantha, Dungapur, Banswara and Banaskantha. About 11.5 percent of the total tribal population of the country is found in this region.

The present study is located in the East Central tribal belt of the country. The economy of the tribes in this region is conditioned by three different ecological settings. These are (i) hills having forest covers, where the tribals practise mainly shifting cultivation of the slash and burn type, and also adopt terraced cultivation in suitable places; (ii) the hills almost completely denuded of forest covers where tribals cultivate such hill and foothill lands with whatever soil cover is left and (iii) the plateaus and plains, where settled agriculture is practised. In addition to agriculture and collection of forest produce, the tribals depend upon agricultural and non-agricultural wage labour where such opportunities are available.

Koraput district (Undivided)* in Orissa has been selected for our study. It is a very large district, covering an area of 27,020 sq. km. which from 17.3 percent of states total geographical area. Situated in the south west corners of Orissa, it is essentially an upland plateau region with spurs of the eastern Ghats cutting it. The three distinct ecological settings that found in the east central tribal belt of the country mentioned above, are also found in this district. Koraput district (Undivided) geographically can be divided into four divisions, viz; the 3,000 ft plateau, which mainly covers the sadar sub-division of Koraput, the 2,000 ft plateau which covers the Jeypore and Nawarngapur sub-divisions, the 1,000 ft plateau covering Rayagada and Gunupur sub-divisions, and the 500 ft plateau which covers the entire Malkangiri sub-division. Koraput is basically a tribal district; 56 percent of its population is tribal, and 52 types of tribes, out of the 62 found in Orissa, are found in this district. Three tribal villages

* Since 1992, Koraput (District) has been divided into four districts (parts) i.e. (i) Koraput, consisting of Koraput and Jeypore Sub-divisions, (ii) Nawarangpur, consisting of Nawarangpur Sub-Division, (iii) Malkangiri, consisting of Malkangiri sub-division, and (iv) Rayagada, consisting of Rayagada and Gunupur sub-divisions. The present study was undertaken in 1983, 1984 and 1985 prior to bifurcation of the districts. Hence, here Koraput district would mean undivided Koraput district.

of the district, located in the three ecological settings, viz., (a) the hills with forest cover, (b) hills without forest cover, and (c) the plateau and plains, have been surveyed for study of tribal economy in depth. In addition, another village (rather a colony), where mainly the tribals who were practising shifting cultivation and were shifted and resettled under the land colonisation scheme in order to examine the success or otherwise of such measure.

THE STUDIED VILLAGES

Sanabaleswar: Sanabaleswar, located in the one thousand feet plateau under Rayagada sub-division with hills around, covered by some type of forest or the others, is typical of the tribal villages of this ecological region where more than 90 percent of the villages have a population of less than 500 each, and more than 60 percent have less than 200 each. Most of the villages like Sanabaleswar have little flat land fit for agriculture. The surrounding hills support some shifting cultivation, heavily supplemented by collection of minor forest produce including bamboo clandestinely, as well as food in the form of roots, fruits and leaves.

Uppergodala: Village Uppergodala is located on the 3000ft plateau where the villages with 200 to 500 persons are relatively more numerous, constituting 35 percent of all villages in the region; smaller ones, with less than 200 people constitute 44 percent of all villages. The relatively large villages are often such because the forest lands in the neighbourhood are bereft of any vegetation and are routinely put to agriculture. The hills slopes are comparatively gentle here than at Samabaleswara, besides some foot hill lands.

Kangumajhiguda : The third village, Kangumajhiguda situated in the 2,000 ft plateau under Nawarangapur sub-division, is a rather large village with 344 people. Most of the villages in this zone are medium and large villages; villagess with population 200 to 499 constitute 31 percent and those with more than 500, nearly 50 percent of the total number of the villages. Being essentially a flat settled agricultural village, there is hardly anything to distinguish such tribal village from similar non-tribal ones in this plateau.

Kumbharadhamuni Adivasi Colony—This colony is located in the eastern part of the 1,000 ft plateau under Gunupur sub-division. Originally 32 shifting cultivator families brought from several villages were settled under a land colonisation scheme during the second five year plan period. Each family was given 3 acres of reclaimed land near the colony for cultivation, with plough animals and other implements as well as inputs at the initial stage. By the time the survey was undertaken (1984-85) only 14 (44 percent) of the original settlers, 5 extended families of these families, and 4 new comers, a total of 23 households, were living in the colony.

Household and Population

Table 1 shows the total number of households in each village, classified into landless and landed and also the number of families taken in the sample for detailed study. The sampling of households was mainly based on landless and land holding (i.e., land put under cultivation during 1983-84) reported in the general households schedule. But due to wide variation in the number of households, their composition and land holding pattern in the four villages, uniform sampling could not be adopted. Hence it varied.

In Samabaleswar, all the larger land owners (only 3 having more than 2 acres) as well as the only landless one, and 50 percent of smaller land owners were selected at random as the sample. In each of the other three villages, 50 percent of the landless families were selected. From among the landed households, 20 percent in Uppergodala, 25 percent in Kangumajhiguda and 100 percent in Kumbharadhamuni colony were included in the sample.

In Samabaleswar and Kumbharadhamuni colony all the household (26 and 23, respectively) belonged to Kondh-tribe, whereas in the other two villages, some non-tribal household were found living amongst the majority tribal families. For example, in Uppergodala out of 94 families 70 belonged to "Paraja" tribe, 11 scheduled caste 'Dombo' and 13 were caste Hindu "Sundhis". Similarly in Kangumajhiguda, out of 79 families, 53 were Scheduled Tribe (45 Pengu Parajas and 8 Bhataras), and rest caste Hindus: 2 Brahmins, 6 Kshatriyas and 10 Gouda families.

Population

The total population of the villages with break up by age and sex is given in Table 2, and the same information for the sample households, in Table 3. Children in the age group 0-14, constituted between 38 and 40 per cent of the total population in three of the

TABLE 1

Number of Families Classified by Landless and Landed, in four Surveyed Villages of Koraput District, During 1983-84

Sl. No.	Type of households	*Villages*							
		Samabaleswar		*Uppergodala*		*Kangumajhiguda*		*Kumbhardhamuni*	
		No. of households	*No. in the sample*	*No. of households*	*No. in the sample*	*No. of households*	*No. in the sample*	*No. of households*	*No. in the sample*
1.	Landless	1	1	10	5	20	10	6	3
2.	Landed	25	14	84	17	59	15	17	17
3.	Total	26	15	94	22	79	25	23	20

Source: General Household Survey, 1983-84.

four villages, the same as in the district (rural) population as per the 1981 Census. But it was only 27 per cent in Kumbhardhamuni Adivasi colony, mainly due to the very small number of children in the age group 0-5. Enquiry with settlers revealed that the male members of almost all the eligible couples, after one or two issues, had undergone sterilization operation. The percentage of population in the age group 15-59, which constituted the principal work force, varied between 51 to 64 per cents in different villages as against 55.7 per cent in the district (rural). Consequently, the percentage of older persons varied from 3.4 to 8.9, as against 4.7 in the district (rural).

Working Population and Productive Activities

The 1981 Census defined a "worker" as one who worked in any gainful activity during the year, a "main worker" as one who worked 183 days or more, and a "marginal worker" as one who worked less than that. The same definition is used here. Detailed data were collected about the daiy engagements of the members of the sample households during the 365 days of the year beginning July 1, 1984. In addition to the daily employment schedule, the economic enterprise schedule sought information on area of different types of land owned and cultivated, crops grown, inputs used and the output raised, capital assets and investment, etc. The activities like work on own farm, wage work on others' farms as casual ad attached labour, other non-farm productive works, livestock work, collection of minor forest produce including firewood, and food gathering were recorded as productive works. On the other hand, tasks like visits to relations and market places, domestic works, sickness, festivity, visits outside the village and time spent in unemployment were treated as unproductive. Table 4 gives the percentage of workers in different age and sex groups in the four villages and Table 5 in the district (rural). Comparison of the percentages of the working population in the four surveyed villages with that for the district (rural) shows that it was high not only for the adult males but also for adult women and children (of 6 to 14 years of age) in the surveyed villages. Among the rural population of Koraput district according to the 1981 census, the workers formed 49.9 per cent, and among the tribal population 53 per cent. But in the four surveyed villages this percentage was 67 in Uppergodala, 73 in Samabaleswar, 76.7 in Kangumajhiguda and 89.2 in Kumbhardhamuni Adivasi Colony. This large difference between the census percentage of workers and that in the surveyed villages is mainly

TABLE 2

Distribution of Total Population of the 4 Surveyed Villages of Koraput District Classified by Age and Sex, During 1983-84

Age Group	Villages											
	Samabaleswar			*Uppergodala*			*Kangumajhiguda*			*Kumbhardhamuni*		
	M	*F*	*Total*	*M*	*F*	*Total*	*M*	*F*	*Total*	*M*	*F*	*Total*
0-5	7	12	19	43	37	80	29	34	63	2	3	5
	(12.5)	(20.0)	(16.8)	(20.4)	(17.6)	(19.0)	(16.9)	(19.7)	(18.4)	(3.8)	(6.0)	(4.8)
6-14	17	13	30	40	40	80	43	44	87	13	10	23
	(30.3)	(21.6)	(25.8)	(19.0)	(19.0)	(19.0)	(25.0)	(25.6)	(25.3)	(24.5)	(20.0)	(22.3)
0-14	24	25	49	83	77	160	72	78	150	15	13	28
	(42.8)	(41.6)	(42.6)	(39.4)	(36.6)	(38.0)	(40.9)	(45.3)	(43.7)	(28.3)	(26.0)	(27.4)
15-59	30	33	63	110	115	225	89	86	175	34	32	66
	(53.6)	(55.0)	(54.0)	(52.1)	(54.7)	(53.4)	(51.7)	(50.0)	(50.8)	(64.1)	(64.0)	(64.0)
60 &	2	2	4	18	18	36	11	8	19	4	5	9
above	(3.6)	(3.4)	(3.4)	(8.5)	(8.7)	(8.6)	(6.4)	(4.7)	(5.5)	(7.6)	(10.0)	(8.9)
Total	56	60	116	211	210	421	172	172	344	53	50	103
	(100)	(100)	(100)	(100)	(100)	(100)	(100)	(100)	(100)	(100)	(100)	(100)

TABLE 3

Distribution of Population Belonging to Sample Households of the 4 Surveyed Villages of Koraput District Classified by Age and Sex

Age Group	Villages											
	Samabaleswar			*Uppergodala*			*Kangumajhiguda*			*Kumbhardhamuni*		
	M	*F*	*Total*	*M*	*F*	*Total*	*M*	*F*	*Total*	*M*	*F*	*Total*
0-5	5	9	14	10	7	17	7	3	10	1	3	4
	(14.3)	(23.0)	(18.9)	(20.4)	(15.5)	(18.0)	(13.2)	(6.5)	(10.1)	(2.1)	(6.3)	(4.3)
6-14	9	8	17	10	9	19	16	14	30	11	10	21
	(25.7)	(20.5)	(22.9)	(20.4)	(20.0)	(20.2)	(30.1)	(3.4)	(30.3)	(23.9)	(21.2)	(22.5)
0-14	14	17	31	20	16	36	23	17	40	12	13	25
	(40.0)	(43.5)	(41.8)	(40.8)	(35.5)	(38.2)	(43.3)	(36.9)	(40.4)	(26.0)	(27.5)	(26.8)
15-59	19	20	39	26	25	51	28	28	56	30	29	59
	(54.3)	(51.3)	(52.7)	(53.0)	(55.5)	(54.2)	(52.8)	(60.8)	(56.6)	(65.2)	(61.7)	(63.4)
60 &	2	2	4	3	4	7	2	1	3	4	5	9
above	(5.7)	(5.2)	(5.5)	(6.2)	(9.0)	(7.6)	(3.9)	(8.3)	(3.0)	(8.8)	(10.8)	(9.8)
Total	35	39	74	49	45	94	53	46	99	46	47	93
	(100)	(100)	(100)	(100)	(100)	(100)	(100)	(100)	(100)	(100)	(100)	(100)

TABLE 4

Percentage Distribution of Workers Classified by Age and Sex in 4 Surveyed Villages, During 1984-85

Age Group	*Villages*											
	Samabaleswar			*Uppergodala*			*Kangumajhiguda*			*Kumbhardhamuni*		
	M	*F*	*Total*	*M*	*F*	*Total*	*M*	*F*	*Total*	*M*	*F*	*Total*
0-5	—	—	—	—	—	—	—	—	—	—	—	—
6-14	77.7	62.5	70.5	30.0	44.4	36.8	62.5	64.2	63.3	81.8	60.0	71.4
0-14	50.0	29.4	38.7	15.0	25.0	19.5	42.4	52.9	47.5	75.0	46.1	60.0
15-59	100.0	100.0	100.0	100.0	92.0	96.0	92.8	100.0	92.8	100.0	100.0	100.0
60 & above	100.0	50.0	75.0	100.0	100.0	100.0	100.0	100.0	100.0	100.0	100.0	100.0
Total	80.0	66.6	72.9	65.3	68.8	67.0	71.6	82.6	76.7	93.4	85.1	89.2

TABLE 5

Percentage Distribution of Workers Classified Age and Sex in Koraput District (Rural) As per the 1981 Census

Age group	*Male*	*Female*	*Total*
0-14	16.3	11.4	13.8
15-59	91.0	50.9	73.7
60 & above	70.5	21.9	44.8
Total	63.3	36.6	19.9
Total S.T. (Rural) Population	65.4	40.8	53.0

Notes: (1) The Census, 1981 gave 0-14 age group as a single unit, without further breakup.

(2) Breakup of the S.T. rural tribals into various age groups was not available.

due to the high percentage of female and child workers. The census (1981) underenumerated the participation of women and children in tribal economy in general, and their role in food gathering and the collection of minor forest produce in particular. In the total rural population of the district only 50.9 per cent of adult women, 15-59 years of age, were reported as workers (both main and marginal). As against this, in three of the four villages surveyed this was 100 per cent; only in Uppergodala it was 92 per cent. The percentage of children of less than 15 years of age (essentially children of 6 to 14 years) who were workers, was only 13.8 for the rural district population according to the 1981 census. In the three of the surveyed villages this percentage was between 40 to 60 (for children of age 6 to 14, who formed about half of all children below 15 years of age); only in Uppergodala it was just below 20 per cent.

Time Spent in Gainful Activities

Not only adult men, women and children in tribal societies participated in large numbers in economically gainful activities, but also (as it will be evident from Table 6) they had spent a very large part of their time during the year in such activities. The adult men in the sample villages spent more than 70 per cent and adult women more than 50 per cent of their total time in productive activities. The time spent by the children in gainful activities remained above 50 per

TABLE 6

Percentage Distribution of 365 days of Adult Male, Adult Female and Child Workers of the Four Surveyed Villages in Various Tasks During July 1, 1984 and June 30, 1985

Sl. No.	Name of the Task	Samabaleswar			Uppergodala			Kangumajhiguda			Kumbhardhamuni		
		Adult Male	Adult Female	Children	Adult Male	Adult Female	Children	Adult Male	Adult Female	Children	Adult Male	Adult Female	Children
1	2	3	4	5	6	7	8	9	10	11	12	13	14
1.	Work on own farm	45.5	24.0	14.0	28.6	27.6	13.5	32.2	14.0	13.9	23.8	19.6	4.1
2.	Wage work on others farms	1.5	0.7	—	8.6	12.2	15.2	25.9	21.3	18.5	11.0	39.3	17.5
3.	Attached labour	—	—	—	6.7	—	—	—	—	25.0	25.5	—	20.5
4.	Live stock work	0.4	—	24.5	4.2	0.3	10.8	3.5	3.1	8.6	6.3	0.1	37.9
5.	Non-farm prod. work	3.7	1.6	0.2	7.0	1.7	0.9	8.6	3.8	0.1	11.9	5.2	3.4
6.	Collection of M.F.P. and firewood	9.9	3.8	2.5	8.3	5.4	5.7	6.3	7.6	10.6	7.5	3.9	1.1
7.	Food gathering	11.2	31.6	19.5	3.7	3.6	4.7	-	-	0.5	0.1	1.6	0.2
8.	Total Productive Works	72.2	61.7	60.7	67.1	50.8	50.8	76.5	49.8	76.7	86.1	69.1	84.8

(Contd.)

Table 6 *(Contd.)*

1	2	*3*	*4*	*5*	*6*	*7*	*8*	*9*	*10*	*11*	*12*	*13*	*14*
9.	Repair and maintenance of own house	0.3	0.1	0.1	0.6	—	—	0.2	—	—	0.1	0.1	—
10.	Household work	17.4	31.4	35.0	16.4	38.9	40.7	6.7	41.1	15.7	7.5	26.4	10.5
11.	Visit to relatives	0.5	0.3	0.3	4.4	3.3	3.3	4.8	3.6	0.8	1.3	0.5	0.6
12.	Festivity	3.9	3.8	2.7	5.0	4.7	4.7	2.8	2.9	3.1	2.2	2.1	2.2
13.	Marketing	1.8	1.5	0.5	1.3	0.6	0.4	2.4	0.9	0.05	0.1	0.1	—
14.	Sick	3.9	1.2	0.7	1.7	1.7	0.1	4.3	1.7	3.6	1.6	1.1	1.9
15.	No work/ Looking for work	—	—	—	3.5	—	—	2.3	—	—	1.1	—	—
16.	Total of non-prod. works	27.8	38.3	39.3	32.9	49.2	49.2	23.5	50.2	23.3	13.9	30.3	15.2
	Grand Total	100.0	100.0	100.0	100.0	100.0	100.0	100.0	100.0	100.0	100.0	100.0	100.0

cent and went upto even 84 per cent in some villages. The National Sample Survey showed that, in 1983, in India the males (5 years and above) spent 55.86 per cent of their time in gainful activities and the female, 22.9 per cent. The rural scheduled tribe male spent 61.64 per cent and females 38.38 per cent of their time in productive employment. In Orissa, during the same year, the rural tribal males spent 62.58 per cent and females 32.38 per cent of their time in gainful activities. However, the workers in the sample villages of Koraput worked comparatively more days in gainful activities than the N.S.S. averages for the state as a whole. Of course, the tribal workers had spent a larger part of their year in gainful activity than the non-tribals in these villages.

Returns from Economic Activities

It is against this background of heavy work done by the tribal workers, that one notices the very poor return to their labour. (Here return to labour is measured as the total value added in different economic activities divided by the total number of adult male equivalent days in the activity during the year, In calculating return per family labour day, the value added in the activity net of wages paid is divided by the total number of adult male equivalent days of household labour.) Of course, the returns differ depending upon the ecological settings in which the tribals live.

Agriculture

Agriculture is the most important source of employment, and also of income, in the tribal areas. Cultivation of own farm land, and agricultural wage labour (including, of course, attached labour) claimed 47 per cent of the adult male days in Samabaleswar, 44 per cent in Uppergodala, 58 per cent in Kangumajhiguda, and 60 per cent of such time in Kumbhardhamuni colony. (Ref. Table 6). Agriculture in tribal areas is carried out both on the hills and on the plateaus and plains.

Samabaleswar: The tribals under Samabaleswar situation practise shifting cultivation in the hill lands over which they have no legal right. As there was no ground measure of land under "Podu" (shifting cultivation), we had to rely on the seed measure, to ascertain the area of land put under cultivation by the tribal farmers. Accordingly, it was found that 63 acres of land were under occupation in the village which included actual area under cultivation during the year and area left fallow for regeneration. About 36 to 53 per cent of Podu land was annually put under cultivation in the village. There were only 7.6 acres of foot hill low dongar-land available for cultivation. Land was fairly equally distributed: only one household had 5 acres and another was landless that year; the rest had 2 acres or less land each.

Mixed cropping is the usual practice under such "slash and burn" (shifting) type of cultivation. Seeds of crops like ragi, kandul (arhar), ganthi, kosla, and suan are generally mixed and sown. Crops like niger and maize are sown as pure crops. A variety of small paddy, called "Dongar Dhan" (hill paddy), is sown on Podu land. Paddy is also grown in terraced fields; but such fields do not exist in Samabaleswar. The sample households in Samabaleswar had put 27.05 acres of Podu land and 5.5 acres of low dongar land under cultivation during the 1984-85 cropping season.

Uppergodala: The farmers of Uppergodala had put 721 acres of land to cultivation during the 1983-84 cropping season, which consisted of 168 acres of encroached hill lands (used for shifting cultivation in the past, but now put to dry cultivation every year as there is no more forest growth on such hills), foot hill dongar land of 373.61 acres, 151.05 acres of terraced paddy land, and 28.75 acres of garden land. There was considerable inequality in the distribution of land in the village. Only 7 highest land holders, all tribals, who had 20 acres or more of land each, possessed 35 per cent of the total land. The majority of the landless families belonged to scheduled caste "Doms", who mainly traded in crops and cattle. Out of the total available land, the 22 sample households had put 100.63 acres of land under cultivation during the year under intensive survey (1984-85), which consisted of 37.90 acres of encroached hill land, 36 acres of low dongar land, 19.65 acres of terraced paddy land, and 7.08 acres of garden land.

In the encroached hill and low dongar lands, the farmers of Uppergodala had sown coarse cereals like ragi, suan, bhodei, oil seeds like alsi (niger), and pulses like kandul (arhar) and chana (gram), etc. Dongar paddy was also grown in these lands. Paddy was grown in terraced paddy fields, and crops like maize, garlic, chilli, onion and some vegetables were grown in the garden lands. The terraced paddy fields get ample water during the rainy season. The water of a small hill stream was being diverted to the village and was used to irrigate only the garden lands. Plough was used in the hills having gentle slope, in foot hill lands, and in terraced fields. The only land that was manured was the garden land.

Kangumajhiguda: The paddy fields are on flat lands specially developed for paddy cultivation with field bunds on all four sides. There are other patches of flat land, not as developed as the paddy fields, which are locally known as dongar lands. A portion of the second type of land has been encroached from the forests, mostly from unreserved forests called "Patra Jungle" over which the people have no occupancy right. Dongar paddy, ragi, blackgram, niger, kolthi, jahna and arhar were the crops grown in such lands. Paddy, of course, was grown in the paddy land, but a few farmers were also found growing wheat, potatoes, garlic, and some vegetables in paddy fields, after paddy was harvested, by irrigating such crops from tanks by manual labour.

The total amount of land under cultivation, as ascertained from the households, was 269.66 acres which consisted of 133.45 acres of settled paddy land, 67.70 acres of settled dongar land, and 68.51 acres of encroached forest land. There was great inequality in the distribution of land in the village. Out of the 20 landless families, 10 belonged to scheduled tribes and 10 to caste Hindus. Only 2.6 per cent of the households with more than 20 acres each (all STs) owned 16.7 per-cent of the land, 12.7 per cent of the households with more than 10 acres each owned 57 per cent of the land; 19 per cent of the households with more than 5 acres each owned 67 per cent of the total land. While a quarter was landless, another 35.4 per cent owned less than 2.5 acres each accounting for only 12.4 per cent of the land, and 55.6 per cent of the households with less than 5 acres each owned 33 per cent of the total farm land. The bulk (62.2 per cent) of the land with the non-tribals was enccroached dongar land.

Kumbhardhamuni Adivasi Colony: Thirty two tribal households were persuaded to leave their villages and settle in this colony during years 1953-58, to practise settled agriculture. Each household was provided with 3 acres of land making a total of 96 acres. By 1984, the 14 original settlers still in the colony had 42 acres of paddy land under their legal ownership. The ownership of the remaining 54 acres of paddy land still stood in the names of the 18 households who deserted the colony. However, the revenue inspector allowed any other household, old or newcomer, which was prepared to pay the legal cess on any land belonging to these deserting households, to occupy such land for cultivation, without of course, the legal ownership on such lands. Fortyfour of the 54 acres of such land was thus occupied. But all this land was not put under cultivation . During the 1984-85 crop season the colony households cultivated only 23.83 acres of paddy land, and leased out 12 acres to non-residents. In addition to this, they had put 12.84 acres of Podu land under cultivation, giving a total of 36.67 acres of land under cultivation. Thus, nearly 73 per cent of the colony land was not being cultivated by the settlers, the bulk of it lying fallow. The settlers stated that the land supplied to them by the Government was, by and large, saline land with very poor yield.

Economics of Cultivation

While calculating the cost of production of crops only paid out costs of hired human labour, hired bullock labour, seed cost—whether home supplied, bought or borrowed—and cost of purchased fertiliser or manure were-taken into account. Imputed cost of family labour and bullock labour, cost of home supplied manure, and land cess as well as depreciation cost of agricultural implements were not included as cost; nor was home supplied manure calculated as return. While the net incomes (net of the above paid out costs) have been calculated for the individual crops, these are not presented here or discussed in detail. Only the overall net returns, in money terms, from different types of land cultivated by the households are analysed here.

In shifting cultivation family labour was the major input, in addition to the seeds. In settled type of agriculture, family labour is supplemented, to a smaller or larger extent, by hired labour. In the Podu lands of Sanabaleswar and Kumbhardhamuni colony which had proper Podu lands, labour days engaged per acre were 153 and 183, respectively. The other types of land did not require as many labour days. (Ref. Table 7). The per acre paid out cost, consisting mainly of hired labour cost and value of seed and purchased manure, was the highest in Kangumajhiguda where only settled agriculture was practised. This resulted in greater difference between gross value of farm output and net farm income, unlike in case of Podu.

The overall performance of agriculture was the poorest in Sanabaleswar, where essentially shifting culativation was practised. Both the gross value of output and the net farm income per acre in this village were the lowest of all. Even the money return per acre of the Podu land in Kumbhardhamuni colony was 2.5 times that in Sanabaleswar. Mixed cropping was the normal practice on Podu lands.

TABLE 7

Per Acre Family Labour Days, Hired Labour Days, Bullock Labour Days and Gross Value of output and Net Farm Income From different Types of Land in Four Surveyed Villages of Koraput District During 1984-85

Sl. No.	*Name of the Village*	*Type of Land*	*Area sown*	*Family lab. days*	*Hired lab. days*	*Bullock lab. days*	*Gross value of output (Rs.)*	*Net farm income (Rs.)*
1.	Samabaleswar	Podu Dongar	32.50	153.0	3.0	4.5	79.80	66.10
2.	Upper-godala	(a) Encroached hill land	37.90	40.0	2.0	8.0	150.29	121.47
		(b) Foot hill dongar	36.00	38.4	2.5	9.7	321.26	268.63
		(c) Terraced-paddy land	19.65	27.4	0.2	9.5	223.91	201.73
		(d) Garden land	7.08	36.8	0.4	9.3	349.03	313.17
		Total	1003.63	36.8	1.7	9.0	239.83	203.14
3	Kangu-majhiguda	(a) Settled Dongar	52.40	49.0	25.4	13.8	452.45	267.63
		(b) Paddy land	40.72	36.2	56.6	14.7	1090.25	729.85
		Total	93.12	43.4	39.0	14.2	731.35	469.64
4.	Kumbhar-dhamuni	(a) Podu Dongar	12.84	18.3	—	2.6	201.25	184.91
		(b) Low Dongar (Bara) land	2.97	133.5	—	27.9	877.43	849.65
		(c) Paddy land	20.86	71.5	6.8	15.4	460.18	365.53
		Total	36.67	115.7	4.0	12.0	403.30	341.49

Not every type of seed sown by the farmer, however, yielded a return; different crops failed in case of different farmers. This appears to provide justification for mixed cropping. Net farm income was the second poorest in Uppergodala. Here also, the poorest land was the encroached hill-land, on which once upon a time there was shifting cultivation. The low dongar land and paddy land here had somewhat better returns, though the best return from paddy lands was in Kangumajhiguda with settled paddy cultivation on flat lands. The garden land, very limited in area, had the best returns in the villages where it was available. This was because more care was taken of these small plots, located adjacent to the house site, and generally fenced, manured and irrigated. But availability of such lands is limited everywhere.

Labour Productivity in Agriculture

How far was labour productive in tribal agriculture? An attempt is made to find out the per labour day return (return to an adult male day, after converting female and child labour days to adult male units) from agriculture in each type of land and in each village, and to compare the same with the prevailing casual wage rate in agriculture in the village during the year under study. The average agricultural wage was Rs. 3 for an adult male worker in Sanabaleswar and Uppergodala villages and Rs. 6 in Kangumajhiguda and Kumbhardhamuni during 1984-85. Labour productivity was the lowest in Sanabaleswar, only Re. 0.42 for family labour and Re. 0.46 for family plus hired labour (Ref. Table 8); it was less than one-sixth of the prevailing daily wage rate. The labour productivity on own farm in Kumbhardhamuni Adviasi Colony was also very poor—less than half the prevailing causal agricultural wage rate in and around the village. It was comparatively better in Uppergodala where labour productivity in agriculture was Rs. 4.32 more than the local wage rate (Rs. 3) in agriculture. In Kangumajhiguda the labour productivity was the best of all the 4 surveyed villages. As for returns to labour day in different types of lands, the poorest are of Podu lands of Sanabaleswar and Kumbhardhamuni, and the encroached hill lands of Uppergodala. Only in the paddy lands in Kangumajhiguda was the return to labour much higher than the prevailing wage rate. In the other types of land in Uppergodala labour productivity was also higher than the prevailing wage rate. But in the Kumbhardhamuni colony returns to labour on paddy and gardens lands were significantly lower than the prevailing wage rate.

Other Gainful Activities

In addition to cultivation of own farms, the tribals pursued other gainful activities, depending upon the ecological situation and the distribution of land holdings. In Sanabaleswar—where Podu lands, fairly equally distributed, had very low productivity—with little opportunity of wage work in the vicinity, food gathering and collection of minor forest produce were the more important activities. In

TABLE 8

Returns per day of Family Labour and Total (Family Plus Hired) Labour from Agriculture in different Types of Lands in the Four Surveyed Villages during 1984-85

(in Rs.)

Sl. No.	*Name of the villages*	*Type of land*	*Return per family labour day*	*Return per hired and family labour day*
1.	Samabaleswar	Podu	0.42	0.46
2.	Uppergodala	Encroached hill land	2.43	2.53
		Foot-hill dongar	5.32	5.27
		Terraced paddy land	5.62	5.62
		Garden land	6.43	6.44
		All land	4.32	4.33
3.	Kangumajhiguda	Settled dongar land	5.46	5.17
		Paddy land	20.22	11.00
		All land	9.96	7.83
4.	Kumbhardhamuni	Podu	0.99	1.02
		Paddy land	4.70	4.81
		Garden (Bara) land	5.43	5.43
		All land	2.72	2.83

Kangumajhiguda and Uppergodala villages, particularly in the former, land was very unequally distributed. There was little forest in the neighbourhood fit for food gathering and collection of minor forest produce; the reserve forest in the vicinity of Kangumajhiguda was useful only for collection of certain types of minor forest produce. Therefore, agricultural wage labour was very significant in both these villages. Similarly, agricultural wage work including attached labour was a more important economic activity than cultivation of own farm in Kumbhardhamuni colony, because of the poor productivity of the paddy land and availability of wage work on farms in the adjacent villages.

The proportion of productive labour days spent in various economic activities, the proportion of income earned from each, and the return per labour day (family labour day) in each village surveyed are given in Table 9. It shows the relative importance of the various economic pursuits both from the employment and income points of view. That the importance of forest collection and food gathering activities in Samabaleswar was far greater than that of agriculture (including agricultural wage work) is seen from the fact that the former two activities occupied 39 per cent of the productive labour days of the workers but contributed more than 74 per cent of total income, whereas the latter two activities, mainly cultivation, occupying 51.4 per cent labour days, but contributed only 17.2 per cent of the total income. Non-farm productive activity seems to be the most productive one in Uppergodala, in which by spending only about 8 per cent of productive labour days, people received 21.6 per cent of the total income (next in importance to cultivation), with the highest return per labour day. This, however, was unusual because it was mainly due to the salary of one tribal who worked as a postal runner and also earned some profit from contract work in road construction. Agriculture, including wage work in it, was the most important occupation, accounting for 69 percent of the labour days, one-third of which was wage labour, accounting for the same share in total income. Collection of minor forest produce and food gathering, particularly the latter, were of minor importance. Kangumahiguda being basically a settled agricultural village, the contribution of agriculture (cultivation and agricultural wage) was very significant: these occupied a little more than 69 per cent of their gainfully employed time and contributed 76 per cent of the total income. Thanks to the very unequal distribution of land, wage labour accounted for the larger part of the time nearly 40 percent, but the smaller part of the income—about 31 per cent. No time was spent on gathering food in the forest; only one-ninth of the gainfully employed time was spent in gathering certain minor forest products which contributed one-eleventh of the total income. It is interesting, however to note that the return per labour day in this activity was higher than in wage work on farm, mainly because of certain special types of minor produce like sal seeds and leaves for plates which fetched good price in the market. In Kumbhardhamuni colony, on the other hand, wage work on others farms occupied almost half (48.2 per cent) of the gainfully employed time and nearly 57 percent of the total income. Indeed, it fetched the

TABLE 9

Percentage of the Days Gainfully Worked, Percentage of Income Received and Income Received per Labour Day from Various Gainful Activities in Four Surveyed Villages of Koraput District During 1984-85

Sl. No.	Name of Activity	*Samabaleswar*			*Uppergodala*			*Kangumajhiguda*			*Kumbhardhamuni*		
		% of days worked	*% of income received*	*Return per lab. day (Rs.)*	*% of days worked*	*% of income received*	*Return per lab. day (Rs.)*	*% of days worked*	*% of income recived*	*Return per lab. day (Rs.)*	*% of days worked*	*% of income received*	*Return per lab. day (Rs.)*
1.	Culti. of own farm	49.9	14.2	0.42	45.4	53.8	4.61	29.5	45.2	9.96	23.3	15.7	3.26
2.	Work on others' farms	1.5	3.0	3.00	23.5	15.5	2.58	39.9	30.8	5.05	48.2	56.5	5.20
3.	Livestock work	6.1	1.2	0.30	5.2	4.0	3.00	11.5	3.5	3.95	9.9	1.0	-
4.	Non-farm production work	3.6	7.4	3.04	7.9	21.6	10.63	8.1	11.4	9.36	10.9	18.2	6.74
5.	Collection of MFP and firewood	10.0	48.8	7.22	11.8	3.6	1.18	11.0	9.1	5.38	7.0	7.9	4.61
6.	Food gathering	28.9	25.4	1.30	6.2	1.5	0.89	—	—	—	0.7	0.7	2.33
	Total	100.00	100.00	100.00	100.00	100.00	100.00	100.00	100.00	100.00	100.00	100.00	100.00

highest return per labour day, next only to non-farm wage work. On the other hand, with 23.3 per cent time spent in cultivation, people received only 15.7 per cent of the total income; if time spent in looking after cattle is included in this, it will account for one-third of the total labour time and only 16.7 per cent of total income. The non-farm productive activities which turned out to be the most, rewarding, however, were of a temporary character: the construction of a new road in the locality in which people got wage employment for some days in the lean season.

Seasonality in Employment

Data were collected about the time disposition during the all the 365 days beginning July 1, 1984 for all male, female and child workers in the studied villages. These were appropriately tabulated to show the seasonality pattern in employment over the year.

The data (not presented here for reason of space) show that the adult male and female workers are heavily engaged in agricultural activities like ploughing, sowing and weeding, etc. from the last week of June to the end of September. These activities occupy between 60 and more than 80 per cent of their time during this period. Then it continues at a slightly lower pitch till the harvesting is over by early December. The activities of collection of forest products and food gathering, which are at a very low pitch in the main agricultural time, increase as soon as rains cease and farm work declines. This is also the time when workers, particularly the female workers, report greater time spent on "household works", a part of which is sure to be disguisedly unemployed time. The productive work of child workers was mostly watching cattle, though some of the children helped their parents in agriculture and, in villages like Sanabaleswar, in food gathering and collection of forest produce. The data on seasonality of employment make it clear that the relatively larger time spent by the tribals, compared to non-tribals, in gainful employment is due to self employment in food gathering and collection of minor forest produce as well as preparation of the forest land in the dry season for shifting cultivation.

Scope for Re-allocation of Resources

In view of the very different rates of return to labour in the different activities pursued by the tribal households, a question might arise: would not the economic condition of the households improve by their allocating more labour time to such occupations so as to fetch higher returns per ady of work? For example, would not the tribals in Sanabaleswar be better off by devoting more labour to collection of minor forest produce, supplemented by food-gathering in the forest, from both of which they earn higher returns per day than from agricuture ? In fact, this is not possible. The rainy season is not suitable for digging in the forest for roots for food, or for cutting bamboo or other minor forest products. The forest at this time should, as far as possible, be left alone so that these can grow properly for later harvesting. And rainy season is the proper time for farming. As was noted above, this is fully reflected in the data on the seasonal pattern of employment. Moreover, exploitation of the forest for income is to-day largely an illegal act. Result: the forest area to be exploited gets unavoidably limited, and there is little time for regeneration, and no re-plantation. In a village like Uppergodala the encroached hill land put to agriculture is poor yielding. But, putting it to other uses, like growing useful trees and bamboo, which would fetch more, requires proper rights on the land, technical help, and alternative source of subsistence during the period the trees will take to be ready for the first round of harvest. Even in a colony like Kumbhardhamuni where the settlers have again taken to Podu in the hills, the reasons are that the paddy land is very poor, and the bulk of the labour for Podu is applied in the dry season when there is not enough alternative wage employment available. All this goes to show that the tribal, given the ecological setting, the institutional arrangement and his knowledge, is making the best allocation of the resources and time at his disposal.

Total Income of the Household

The incomes earned by the sample households in different villages under the differing ecological settings are given in Table 10. Sanabaleswar, with its poor resources base, had the lowest level of income—Rs. 200 per capita—among all the four surveyed villages, the highest being in Kangumajhiguda—Rs. 915.21 per capita. It was Rs. 468.39 in Uppergodala and Rs. 864.25 in Kumbhardhamuni colony. How poor were these earnings? The Ministry of Rural Development, Government of India, in order to select the poorest households to be eligible for IRDP benefits, kept the yardstick of Rs. 3,500 per family or Rs. 700 per capita (with base 1979-80). Any family or individual not able to earn this income would be treated as poor. This yardstick

TABLE 10

Total Income Earned by the Sample Households in Four Surveyed Villages of Koraput District from All Sources, Income Received per Household and per capita during 1984-85

Sl. No.	*Name of the village*	*No. of Sample households*	*No. of persons*	*Total income recorded (Rs.)*	*Income per household (Rs.)*	*Per capita income (Rs.)*
1.	Samabaleswar	15	74	14,808.00	987.20	200.00
2.	Uppergodala	22	94	44,029.00	2,001.31	468.39
3.	Kangumajhiguda	25	99	90,606.22	3,624.24	915.21
4.	Kumbhardhamuni Adibasi Colony	20	93	80,375.78	4,018.68	864.25
	Total	82	360	2,29,819.00	2,802.67	638.38

was for the Sixth Five Year Plan and our data relate to the last year of the Sixth Plan, i.e. 1984-85. Even by this measure, unadjusted for the rise in the price level since 1979-80, the average level of living in Sanabaleswar and Uppergodala was below the poverty line. By expressing the average income of the households in 1984-85 at 1979-80 prices (with the help of consumer price index for agricultural labourers in Orissa), we find that the average household income in Kangumajhiguda was Rs. 2,609.45 (per capita Rs. 658.80) and in Kumbhardhamuni colony Rs. 2,893 (per capita Rs. 622). In fact, it shows that the bulk of the households in the four surveyed villages lived below the poverty line.

Approaches to Development

In the hill regions lilke Sanabaleswar, where little land fit for agriculture is available, a more productive method of land use would be forest farming, which is indicated by the high return per day of collection of minor forest produce, mainly bamboo. Indeed, by selling a head load of stolen bamboo once every month for about ten months in the year, the average tribal household in this village earned almost half its total monthly income from all sources. Allocating adjacent forest land to the tribals for forest farming, mainly for bamboo and fuelwood, will not only prevent destruction of forest but also prevent Podu cultivation and provide the tribals with a better source of living. Besides land and the necessary saplings, the state should ensure sustenance through wage employment to the tribal households during the period the trees will take to be ready for the first round of harvesting.

In the ecological setting in which Uppergodala is located, besides some terraced and low lying paddy lands, there is considerable hill top land, once put to Podu cultivation but now annually cropped. These lands are more suitable for growing trees, in the same manner as in Sanabaleswar. Unirrigated horticultural crops can of course be grown on these lands in these two types of villages. But marketing of fruits requires different type of infrastructure and organization which will take quite some time to develop; in the meanwhile the tribals, familiar with marketing timber, firewood and bamboo, can do silviculture. The other lands, fit for agriculture, need the same type of developmental approach as other settled agricultural lands.

Kangumajhiguda, on the other hand, is a tribal village with the same type of settled agriculture and inequality in land distribution as in other non-tribal villages. There is nothing specially 'tribal' about theses villages, except the ethnic composition of the population. These tribal and non-tribal village need a common approach to agricultural development. The reserve forest in the vicinity of this tribal village suggests greater possibility for exploitation of minor forest products than at present. But this depends upon development of proper marketing organization.

The experience in Kumbhardhamuni colony is an object lesson in how not to resettle tribals, and what can wean tribals away from Podu cultivation. The reclaimed lands were poor in quality, with saline soil; and, no efforts were made in all these years to help them with technical information to improve the lands. Fortunately, availability of wage work on farms in the neighbourhood has helped many tribals. If they have again taken to Podu cultivation, it is because the major part of the work for it has to be done in the dry season, when less wage work is available on farms. If other non-farm wage work is available at this time, as was the case with some of the households in this colony, the tribals will automatically turn away from Podu farming. Indeed, this is the lesson for every region where shifting cultivation is still practised. Of course, the alternative opportunity has to be assured regularly and not occassionally.

The study shows that performance of tribal agriculture and returns to tribal labour in different economic activities depend on the quality of land and the ecological setting in which the tribal is set. Given the resource allocation, institutional set up and technology, the tribal is making the best possible use of his resource and time. Improvement in his situation requires change in technology, meaning change in the pattern of land use and all that it involves, and the necessary institutional changes. The pattern and approach will vary with the ecological setting; uniform approach would be inappropriate.

REFERENCES

Baeley, P.G. (1960), *Tribes, Caste and Nation: A Study of Political Activity and Political Change in Highland of Orissa.* London: Oxford University Press.

Behuria, N.C. (1965), *Final Report on the Major Settlement Operations in Koraput District:* (1938-64). Cuttack: Orissa Govt. Press.

Bell, R.C.S. (1945), *Orissa District-Gazetteers,* Koraput Cuttack: Orissa Govt. Press.

Bose, N.G. (1971), *Tribal Life in India,* New Delhi, National Book Trust of India.

Census of India, (1961), *Tribes of Orissa,* Vol. XII, Part V-B: Cuttack, Orissa Govt. Press.

——— (1971), *District Census Handbook,* Koraput, Vol I: Cuttack: Orissa Govt. Press.

——— (1971), *Scheduled Castes and Scheduled Tribes,* India Series 1, Paper 1 of 1975, New Delhi, Govt. of India Press.

Dalton, George (1971), *Economic Anthropology and Development,* New York: Basic Books Inc.

Dandekar, V.M. and N. Rath (1971), *Poverty in India.* Pune, Indian School of Political Economy.

Deogaonkar, G.G. (1980), *Problems of Development of Tribal Areas,* Delhi, Leeladevi Publications.

Dhar, Usha (1970), "Shifting Cultivation in Hill Tracts of India" *IJAE.* Vol. XXV. July-Sept. No. 3.

District-Statistical Office, Koraput, (1970, 1963, 1979, 1980, 1981, 1991 issues) District Statistical Handbook, Koraput.

Elwin, Verrier (1943), *The Aboniginals.* London: Oxford University Press.

———, (1964), *The Tribal World of Verrier Elwin,* London, Oxford University Press.

Fernades, W. and Geeta Menon (1987), *Tribal Women and Forest Economy—Deforestation, Exploitation and Status Change,* New Delhi, Indian Social Institute.

Francis, W. (1907), *Vizagapatam District-Gazetteers,* Govt. of Madras.

Goswami, P.C. (1971) (Ed.) *Socio-Economic Research in Tribal Areas.* Jorhat (Assam). Agro-Economic Research Centre for North-East India.

Ghurye, G.S. (1959), *The Scheduled Tribes,* Bombay, Popular Book Depot.

Govt. of India, *Report of the Commission for Scheduled Castes and Scheduled Tribes,* New Delhi (1959, 1975, 1984, 1985 issues).

——— (1960), Ministry of Home Affairs (1960) *Report of the Special Multi-Purpose Tribal Development Committee,* New Delhi.

———, (1961), Report of the Scheduled Areas and Scheduled Tribes Commission (Chairman U.N. Dhebar), New Delhi.

———, Occasional Papers on Tribal Development, New. Delhi (1976, 1977, 1978, 1979, 1980, 1981, 1988 issues).

Govt. of Orissa, Tribal Research Bureau (1960, 1961, 1964, 1965, 1966, 1970 issues).

Haimendorf, Ch. V.F. (1975), "Pattern of Development for Tribal Societies", *Development Policy and Administration Review,* 1:2, July-Dec.

———, (1982) *Tribes of India—The Struggle for Survival,* New Delhi, Oxford University Press.

Herskovits, M.J. (1974), *Economic Anthropology* New Delhi: Euresia Publishing Company (P.) Ltd.

Mehta, Vaikunth (1959), "Problems of Tribal People". Khadi Gramodyaga.

Mohapatro, P.C. and D. Panda (Ed.), (1978), *Tribal Problems of Today and Tomorrow,* Bhubaneswar, Sabaricultural Society.

Nag, D.S. (1958), *Tribal Economy,* New Delhi, Bharatiya Adimjati Sevak Sangh.

Nanda, P.K., P.C. Mahapatro and J. Sawal (Eds.) (1990), *Communication Barrier in Tribal Development,* Berhampur (Orissa), Das Brothers.

Nevilla, A. Walts (1970), *The Half-Clad Tribes of Eastern India*, New Delhi, Orient Longman.

Patel, M.L. (Ed.) (1972), *Agro-Economic Problems of Tribal India*, Bhopal, Progress Publishers.

Patnaik, N. (1976), Tribes and their Development, Hyderabad, NICD.

Ramaiah, P. (1988), Issues in Tribal Development, Allahabad: Chugh Publications.

Ramamani, V. (1988), *Tribal Economy—Problems and Prospects*, Allahabad, Chugh Publications.

Rao, P. Sasi Bhushana, 1985, Tribals and Forest Ecosystem, P. 151, Konput.

Rath, N. (1985), "Garibi Hatao; can IRDP do it?" *EPW*, Vol. XX. No. 6 Feb. 9.

Rayappa, P.H. and D. Grover (1980), *Employment Planning for Rural Poor—A Case of SCs and STs:*, New Delhi, Sterling Publications.

Sahu, L.N. (1942), *Hill Tribes of Jeypore*. Cuttack, Orissa Mission Press.

Samal, J. (1985), "*Tribal Economy and Forest Ecology,*" P.S.B. Rao (Ed). Tribals and Forest Eco. System, Koraput.

——— (1985), "Tribal Economy—A Regional Study"—Orissa Economic Journal, Vol. XVIII Nos. 1 and 2, Bhubaneswar

——— (1979), Problems of Tribal Agriculture in India, M. Phil Dissertation

Saxena, R.P. (1964), *Tribal Economy in Central India,* Calcutta, Farma K.L. Mukhopadhyay.

Sharma, B.D. (1978), *Tribal Development, The Concept and the Frame*. New Delhi, Prakash Publishers.

Senapati, N. and N.K. Sahu (1966). *Orissa District Gazetteer,* Koraput: Cuttack: Orissa Govt. Press. Ph.D. Thesis.

Mahopatra, P.C. (1980), *Strategy of Economic Development of Tribals of a Backward District in a Backward State—A Regional Study of Economic Problems of Tribals of Koraput District, Orissa,* Doctoral Dissertotion—Utkal University.

Saha, N. (1970), *The Economics of Shifting Cultivation in Assam*. Doctoral Dissertation, Gauhati University.

Upadhyay, A.K. (1982), "*Peasantisation of Adivasis in Thana District, Doctoral Dissertation* GIPE, Poona.

Samal, J. (1992), "*Some Aspects of Tribal Economy—A Case Study of Koraput District*" Doctoral Dissertation, GIPE, Poona University.

18

Environmental Education for 21st Century

NABA KUMAR PATNAIK AND P. SASI BHUSHANA RAO

Life and natural environment are closely related to each other. Without natural environment the creation and survival of life are impossible. The human being, the highest creation of the Author of Nature too benefits from the natural environment. But most of the times man forgets his responsibility in maintaining ecological balance with personal pleasure and gains taking priority. The mad-race after materialism compels man to exploit the natural resources thoughtlessly without taking into consideration of the inevitable consequences relating to the life of the human species of the earth. Considering the importance of environmental education for creating environmental awareness and responsibilities towards environmental protection for ecological balance on the NPE 1986 and 1992 of India, it is stated, "There is paramount need to create a consecciousness of the environment. It must permeate all ages and all sections of society, beginning with the child. Environmental consciousness should inform teaching in schools and colleges. This aspect will be integrated in the entire educational process.

WHAT IS ENVIRONMENTAL EDUCATION?

Environmental education appears to be simplistic and superficial, but actually it is comprehensive and deep. Environmental education is a medium and process of education and that it covers man's relationship with his natural as well as social and man-made environment, and also it includes the relationship of population, industrialization, pollution resource allocation and depletion, conservation, transportation, technology, energy and urban and rural planning to the total biosphere. Viewed so, the crux of environmental education is multidisciplinary in character and its quintessence is a commitment on the part of one and all of us inhabiting this planet earth to prevent deterioration of air, water, land and physical and social environment including inter-relations among people so that a nuclear war, a chemical warfare or any cataclysm generated by man may not destroy the world. Further, environmental education should be a life long process and should aim at not merely imparting knowledge and understanding of man's total environment and of the methods and their application for improving our near and distant surroundings but it should also aim at including skills, the attitudes and values necessary to understand, appreciate and improve our biosphere and troposphere.

OBJECTIVES OF ENVIRONMENTAL EDUCATION

The objectives of environmental education has three domains—cognitive, affective and psychomotor. The objectives in the cognitive domain are:

1. To help acquire knowledge of the immediate environment.
2. To help acquire knowledge of the environment beyond the immediate environment including distant environment.
3. To help understand the biotic and abiotic environment.
4. To help understand the effect of unchecked population growth or unplanned resource utilisation on the world of tomorrow.
5. To examine trends in the growth of population and interpret them for the socio-economic development of the country.
6. To evaluate the utilisation of physical and human resources and suggest remedial measures.
7. To help diagnose the different causes of environmental pollution and to suggest remedial measures.

8. To help diagnose the cause of social tensions and to suggest methods for avoiding them.

Besides the foregoing objectives, the following skills and abilities also fall in the cognitive domain.

1. To help develop observational skills and notice details usually not seen by untrained eye.
2. To help develop skills required for making discriminations in form, shape, habits and habitats.
3. To help develop ability to draw unbiased inferences and conclusions.
4. To help develop ability to make meaningful suggestions.
   The objectives in the affective domain are:
   1. To help acquire interest in the flora and fauna of the near and also distant environment.
   2. To help evince interest in the people and problems of the community and society.
   3. To show tolerance towards different castes, races, religions and culture.
   4. To appreciate the gift of nature.
   5. To love the neighbours and value mankind as a whole.
   6. To value equality, liberty, fraternity, truth and justice.
   7. To respect the national boundaries of all countries.
   8. To value the cleanliness and purity of our environment. The objectives in the Psychomotor domain are:
      1. To participate in afforestation programmes.
      2. To participate in programmes aimed at minimising air, water and noise pollution.
      3. To participate in programmes aimed at preventing soil erosion.
      4. To participate in programmes aimed at eliminating food contamination and adulteration.
      5. To participate in cleaning neighbourhood.
      6. To participate in urban and rural planning and execution programmes, such as installation of gobar gas plants, solar heaters etc.

PRINCIPLES OF ENVIRONMENTAL EDUCATION

The Principles of environmental education are:

1. To consider the environment in its totality viz. economic, political, social, technological, cultural, historical, moral and asesthetic aspects.
2. It should be a continuous process.
3. It should be a life long process beginning from pre-school level to death.
4. To help learners for discovering the real causes of environmental problems.
5. The environmental education should utilise diverse learning environments and broad array of educational approaches to teaching-learning about and from the environment with the stress on practical activities and first hand experience.
6. It should develop critical thinking and problem solving skills of the students.
7. It should stress the complexity of environmental problems.
8. The learners should pay role in planning their own learning experience with others.
9. Opportunities should be provided to the students/learners for making independent decision and face their consequences.
10. The inter-disciplinary approach should be followed.
11. The historical prospectives should be taken into consideration while dealing with present environmental situations.
12. The holistic approach should be adopted.
13. The local environmental issues should be examined alongwith national and international. So that students receive insights in environmental conditions in other graphical areas.
14. The environmental education should explicitly consider environmental aspects in plans for development and growth.
15. The environmental education should relate environmental sensitivity, knolwedge, problem solving skill and value clarifications at every age but with special emphasis on

environmental sensitivity to the learners own community in early years.

SCOPE OF ENVIRONMENTAL EDUCATION

The scope of environmental education means the area ot covers. The following areas are being covered by environmental education.

1. Size and Growth of Human Population

The size and growth of population have positive relationship with the environmental problems. Over population in any part of the world possess tremendous pressure on the natural environment due to over explosion. Similarly the growth of human population directly or indirectly affects the quality of life due to imposibility of expansion of land and other resources.

Day to day the resources—man ratio is increasing due to over exploitation of the natural environment by ever growing population. So the environmentalists always keep an eye on the growth rate of population. Population explosion is one of the major causes of environmental problems and constraints in Planning.

2. Pollution

The environmental pollution, its type viz land pollution, air pollution, water pollution, sound pollution and their negative contribution towards environmental quality deterioration; the broad sectors from which the wastes are produced and above all its effect on the quality of human life and under the purview of discussion in environmental education, the causes and remedial measures of pollution are also being discussed.

3. Human Health and Quality of Life

Human health and quality of life are influenced by the environment to a great extent. In addition to the genetic factors, the determinants of the environment viz. water supply, urban environment, quality, climate and the pattern of human contacts are involved. The pattern of human contacts are involved in production of diseases viz. communicable diseases, degenerative diseases and neo-olastic diseases. In developing countries, these contacts are of major concern. In developing countries malnutrition and various supersition are related to diseases and death. So environmental education issues warnings to the people in this regard.

4. Values, Attitudes and Life Style of Man

Most of the environmental problems of the present day are essentially man made. The role of man is therefore important because it is his attitude and values which shape the environment. In developing countries mad race after materialism puts too much pressure on environment. The materialistic attitude of man multiplies the human needs and ultimately leads towards resources exploitation, without considering the future consequental disharmony with the environment. Man has to evolve a balanced way of thinking, acting and feeling towards the environment. So human ethics, attitude, values, decision making analysis of ecological, economic, social and technical aspects come under the scope of environmental education.

5. Law

Now-a-days it is difficult to cultivate positive attitude among the people towards protection of natural resources. Sometimes people who are being paid for the protection of the resources, spoil it for personal gain. Today people speak one thing and do just in opposite direction due to lack of values. So law relating environment should be framed and observed to such an extent that man does not dare to exploit the natural resources in negative direction. The fear of punishment will change the life style and ultimately the human values and attitudes. Hence education of protection laws are too important for a better environment.

6. Resources

Environmental education covers different type of resources viz. soil resources, plant resources, wild life resource, mineral resources, energy resources, water resources, their sources process involving reduction in the quality as well as quantity resources, problems of resource detorioration, way and means of conserving natural resources, way and means to prevent environmental degradation like crop rotation system, aforestation, use of alternative energy resources such as solar, tidal, wind and geothermal energy and recycling of wastes etc.

FUNCTIONS OF THE ENVIRONMENTAL EDUCATION

The functions of environmental education can best be understood if we identify the people for whom it is required and who should be trained as they only play significant role in the society in creating healthy environment. According to Dr. Pritchard (From the report of the UNESCO Biosphere conference) the following are the different groups of people to whom E.E. helps in various ways for creating healthy environment.

1. It helps the biologist, geographers, geologists, agricultural and forest scientists, farmers and foresters who embark on career on the earth and life sciences.
2. It helps the planners, land scape designers, architects, civil engineers who deal with the design, construction and control of projects affecting the environment.
3. It helps the physicists, chemists and technologists to conduct research without affecting the environmental balance.
4. It helps the stateman, public servants and other leaders to formulate policies keeping in view of the far reaching effects on environment.
5. It helps those who in future will bear the task of interpreting knowledge to young people as educationists.
6. It helps these people who without any direct professional involvement to form a collective voice which will influence other dignitaries of the above categories.

ENVIRONMENTAL EDUCATION FOR PEOPLE IN INDIA

Environmental pollution in India is a serious problem now and serious efforts are made to orient the public in its protection. It became necessary to develop a suitable programme for creating environmental awareness among various sections of society and orient them for adopting environmental friendly production and consumption practices in day to day life.

For that instance, population both in rural and urban areas can be categorised broadly into three groups. These are the children who are attending schools and colleges, illiterates and educated and elite people.

(A) Children

Children and youth upto the age of 20 years, can be inclulded in this category, although there is a wide variation among the school children of urban and rural areas as well as the children of different age groups including the school dropouts. It is easier to approach school children because of their ability to read, write and understand the subject easily. For these children who are illiterate or school dropouts, different communication tools will have to be adopted.

Being fresh and open minded children are anxious to learn. However classroom lessons are ineffective. They can be motivated very easily to initiate suitable activities in the interest of their school community and nation without any vested interest. This is the greatest advantage. Thus for enlisting active participation of children the lesson should be practical and simple enough to be practised by themselves. Good lessons learnt at this stage will certainly turn the children into responsible citizens of the future. The earth can remain safe for long, if we impart environmental education to our children to-day.

(B) Uneducated Adults

A majority of the villagers are or illiterate or semi-illiterate in India. Similarly a large population of women inducting those residing in urban areas are also uneducated. They are not able to understand the gravity of environmental pollution. Unfortunately, a majority of the poor and weaker section of the society fall in this category of semi-literate or iliterates who are struggling to make their livelihood. This is an important section of the target group which is most affected by the pollution. As their primary objective is to earn livelihood, the effective method of motivating them is to link environmental protection with income generation activities.

(C) Educated People

This target group generally consists of well-to-do people with a sound educational background and assured source of income, but limited enthusiasm to take an active part in the field projects. Even then it is necessary to address this group effectively, because of their ability to influence a majority of population as well as the policy makers in the government. This group has to be enlightened with scientific reasoning and logical conclusions about the causes of environmental pollution, hazardous effects and suggested methods to solve the problems. These people can also initiate certain activities in their residential areas or community projects to prevent pollution.

All the three groups need environmental education in order to adopt an eco-friendly life style and to initiate a mass movement for preventing environmental pollution at the regional, national and global level. It is easier to introduce environmental education for children through the schools and colleges. The lessons can be both formal and non-formal. While non-formal education on environment can be introduced immediately with the cooperation of school administration and teachers, it is necessary to approach the State Education Department for including this subjects in the school carricula at various levels.

For illiterate children and adults, it is necessary to adopt non-formal method of environmental education using audio-visual media. Personal contacts are also necessary to introduce this group. As this group is widely scattered, school children, teachers, youth clubs, voluntary agencies and local opinion leaders can influence this group and create awareness among them. Regular use of mass media such as radio, T.V., Cinema, street plays and village meetings are effective. The educated group can be oriented on environmental protection through newspapers, journals, mazazines, radio, T.V. and book-lets apart from public meetings and seminars.

MEDIA AND METHOD

While imparting environmental education and creating awareness, it has been found that the use of multi-media is more effective than restricting to a single medium. The traditional method of formal teaching in the schools may not create any impact on the children. Apart from the test books and other literature, tools such as posters, slides, video films, puppet shows, street plays, field trips, essay and painting competitions and practical projects are also necessary for sustaining the interest of school children. The concept of "do yourself" can have a long lasting effect on them. Village level meetings and field demonstrations are effective in rural areas. Radio and T.V. are other useful media in spreading the messages. Progressive farmers can play a significant role in influencing other members of their community.

There is further scope for involvement of environmental education though well designed training courses, publications and documentary films to orient the children, educated the people, industrial workers, farmers and others for 21st century.

References

Balasubramaniam, Arun-Eco. Development: Towards a Philosophy of Environmental Education—Regional Institute of Higher Education and Development, Singapore, 1984.

Bandhu, D. Singh, H. and Mitra, A.K., Environmental Education and Sustainable Development—Indian Environmental Society, New Delhi, 1990.

Bandhu, D. Bongarti, H. Ghaznawi, A.G. and Gopal, B., Environmental Education for Sustainable Development, Indian Environmental Society, New Delhi, 1994.

Gill, J.S., Environmental Education in the School Curriculum, Developed by NCERT, New Delhi, 1995.

Mohapatro, Damodar, Environmental Education,Kalyani Publisher, Ludhiana, 1995.

NCERT, Inservice Teacher Education Package, (Vols. I & II), 1988.

Trivedi, Priya Ranjan, International Encyclopedia of Ecology and Environment (Vol. 3), Indian Institute of Ecology and Environment, New Delhi, 1994.

Patro, S.N. and Mishra, M.K., Proceedings of the Seminar on Environmental Education Curricula and Environment and Natural Resources Management—Orissa Environmental Society, Bhubaneswar, 1983.

Rao, P. Sasi Bhushan, In Corporate Social Accounting and Reporting, Deep and Deep Publications, New Delhi, 1998.

19

Enforcing Environmental Laws: Some Issues

Krishan Mahajan

It was only in 1977 that environmental protection found a specific mention in the Constitution of India. Twenty eight years after our Constitution came into force, Parliament by the forty second amendment in 1977 made it the primary duty of every government at the Centre and in the States to promote and protect the environment. Article 48A was put into the Directive Principles, which the Constitution states to be fundamental in the governance of the country, to declare: "The State shall endeavour to protect and improve the environment and safeguard the forests and wild life of the country". By the same amendment Article 51A(g) of the Constitution made it a fundamental duty of every citizen to "protect and improve the natural environment includng forests, lakes, rivers and wildlife and to have compassion for living creatures". However it was only in 1980 that the first environmental case came to the Supreme Court, Ratlam Municipality *v.* Virdhichand Jain. Then as the problem of economic development *v.* environment developed the apex court was flooded with petitions relating to pollution of rivers, ground water, air and the reckless

---

Reprinted with permission from Yojana 41:8, August 1997.

destruction of forests. Today the Supreme Court, whose judgments are binding on all under the Constitution, has laid down principles. The problem now is no longer of the existence of legal principles but of their compliance.

The first Supreme Court case of Ratlam Municipality reflected the reality of an India developing into cities with municipalities either unwilling or unable to supply the basic civic amenities for all those within their area. The residents of Ratlam municipality tormented by the stench and stink of open drains, the open spill of an alcohol plant and the public defecation by nearby slum dwellers, asked the Municipality to do its legal duty under the Municipal Act of stopping all this. When this failed the residents moved the local magistrate under the Criminal Procedure Code for prohibition of nuisance. The Magistrate directed the municipality to draft a plan within six months to end the nuisance. The sessions court reversed the order, the high court reversed the sessions court and the municipality came to the Supreme Court in appeal.

Three binding principles were laid down by the Supreme Court. One that no municipality in India can put forth lack of money as a ground for not discharging its primary duty of looking after the health and safety of its residents. Two, the absence of public conveniences and the prevention of industrial pollutants to the detriment of the citizen's health are violations of the human rights of decency and dignity which are non-negotiable first charges on a municipality. Three, officials who violate these duties by not ending the nuisance are subject to punishment of jail and/or fine under the Indian Penal Code. In a telling conclusion the apex court declared: "The officers in charge and even the elected representatives will have to face the penalty of the law if what the Constitution and follow up legislation direct to do are defied or denied wrongfully. The wages of violation is punishment, corporate and personal". However, the judgment made no reference anywhere to the 1977 amendment of the Constitution which put in Articles 48A, and 51A(g). It was only in 1995 that the apex court in Virender Gaur *v.* State of Haryana declared that sanitation and hygienic environment is a fundamental right of citizens and a corresponding constitutional duty of municipalities. The court held that the word 'environment' is of broad spectrum which "brings within its ambit hygienic atmosphere and ecological balance".

In 1974 Parliament had enacted the Water (Prevention and

Control of Pollution) Act. The first case under this Act by way of public interest litigation was decided in 1991 by the Supreme Court. Subhash Kumar *v.* State of Bihar concerned the discharge of coal slurry into the Bokaro river. The apex court put environment on the highest possible legal pedestal by declaring it to be a fundamental right inherent in the fundamental right to life in Article 21 of the Constitution. The citizen's right to clean air and water was thus put beyond the ordinary amending power of ruling politicians. In a simultaneous warning on the possible misuse of public interest litigation in environmental matters, the court declared: "Right to live is a fundamental right under Article 21 of the Constitution and it includes the right of enjoyment of pollution free water and air for full enjoyment of life. If anything impairs or endangers that quality of life in derogation of laws, a citizens has a right to have recourse to Article 32 for removing the pollution of water or air which may be detrimental to the quality of life. A petition under Art. 32 for the prevention of pollution is maintainable at the instance of affected persons or even by a group of social workers or journalists. But recourse to proceedings under Art. 32 of the Constitution should be taken by a person genuinely interested in the protection of society on behalf of the community". By 1993 the Supreme Court gave teeth to this new fundamental right by declaring in Nilabati Behera *v.* State of Orissa that there is a fundamental right to monetary compensation for the violation of a fundamental right. The right to compensation is distinct and separate from the other legal remedies of prolonged civil suits for damages.

TIDAL WAVE

After the creation of compensatory fundamental right through public interest litigation there came a tidal wave of judgments from 1995-97. In Ganesh Wood Products case concerning the Katha forests of Himachal Pradesh the apex court laid down sustainable development as a legal obligation of every government. This meant that before ruling politicians and administrators granted permission for any natural resource based industry they would have to first examine the impact of that industry on the resource base. The impact study must not only relate to current needs but also take into account the needs of the future on the principle of inter-generational equity. The present generation could not eat up all the resources leaving nothing for the future generations. For this purpose the court declared that there was no distinction between public or government forests and private forests since forests as a whole constituted national wealth.

Having laid down four basic principles the apex court went further in Indian Council for Enviro Legal Action (H acid) case Vellore Citizens Forum (tanneries) and S. Jagannathan (aquafarming) cases and laid down three more. One, that it is the polluter who pays. Two, the precautionary principle, namely, that it is the duty of all authorities to prevent pollution or hazards to health of the people. Three, the principle of ecological restoration. This means that the polluter will not only pay for the damage done but also for restoration of the ecology destroyed by him in an area.

The litigative battle to reach these seven principles had been tortuous and full of contradictions as reflected in the Oleum gas leak case of Shri Ram Food and Fertilisers and then that of Union Carbide Corporation. Both cases reflected the tussle between inherently dangerous chemicals rquired for industrial development and the dangers to health or safety of citizens arising from the escape of these chemicals. It is only after these unfortunate incidents, in which the Union Carbide Bhopal factory became the source of the world's worst industrial tragedy, that Parliament enacted the comprehensive Environment (Protection) Act, 1986 and the Public Liability Insurance Act, 1991. The Shri Ram case laid down that "hazardous or inherently dangerous activity for private profit can be tolerated only on condition that the enterprise engaged in such hazardous or inherently dangerous activity for private profit indemnifies all those who suffer on account of the carrying on of such activity regradless of whether it is carried on carefully or not". But shortly thereafter as the methyl isocyanate gas escaped from the Union Carbide Bhopal plant it showed that the problem was not the absence of the laws but their enforcement. When that gas escaped there were a host of laws to carefully regulate such activity under the Factories Act, the Criminal Procedure Code, and the 1981 Air (Prevention and Control of Pollution) Act. The Carbide case resulted in a special law by Pariament by which the then ruling government at the Centre took over the entire litigation for compensation claims on the legal theory of the State being the family father of all its citizens. It also produced a controversial settlement from the apex court which the court itself had to later revise and modify concerning the criminal prosecution exemption granted by it. The criminal litigation still continues as also the activities of machinery of commissioners to process the claims.

GREEN BENCHES

It is in the grid of seven principles, marked by awful contradictions of silence between enactment and enforcement that the apex court has handled Ganga pollution, industrial pollution, the Delhi ridge, automobile pollution, the destruction of forests all over India and the import of toxic chemical wastes. After laying down the principles and ensuring the initiation of some action several of these cases have been transferred to the respective high courts by requesting their Chief Justices to create Green Benches for expeditiously dealing with these matters.

There have been two approaches by the court for implementation of its directives. One is that of calling the officials concerned straight to the court, telling them their legal duty and asking them if they need any orders for any problem they face in discharging this duty. The court has then with their help bound them to a fixed time within which the needful would be done by them. Where the court is not satisfied it has sent its commissioner advocates to monitor compliance and report back to it. The power to appoint commissioners of its own to collect relevant facts in serious breaches of fundamental rights had been declared by the apex court in the 1984 Bandhua Mukti Morcha *v.* Union of India case and then endorsed by a five judge bench of the court in the Shri Ram Foods and Fertiliser (M.C. Mehta *v.* Union of India) case. If performance is found seriously wanting then the court has used its contempt power to secure obedience. The other approach is not to have this face to face dialogue with the officers concerned but simply pass directions on the information given by or through a counsel appointed by the court to help itself. Which is the more successful method to protect the citizens's health and environment? That has not been surveyed anywhere yet. Meanwhile many of the suggestions given by the apex court concerning the duties of ruling politicians in creating an environmentally safe society remain on the pages of judgements. Sadly, the official legal aid machinery has yet to wake up to the need of legally assisting millions of the poor in slums and villages denied both the right to a dignified life and a healthy environment.

SLOW DOWN

The task is huge as the apex court itself realised while hearing the environmental cases and then shifting several of these out to the Green Benches of various high courts. But the overall machinery to undertake the task is woefully lacking. Environmental tribunals, courts have still to come into existence, Speedy remedy, the essence of the relief in such matters, seems to grow more distant. Formerly the Supreme Court used to have a sitting at least once every working week for a full fledged hearing of environmental matters. If necessary daily hearings were held the extent of ecological damage could be decided speedily after hearing all the parties. Officials used to be called and put under a sense of urgency desired by the apex court. All this is no more as the apex court moves on with other legal business.

Two outstanding symbols of this slow down on the legal front are the automobile pollution and the shifting of industries cases. The automobile pollution case concerning Delhi has been going on since 1986. It continues till today with order after order but no final lasting solution. There is no machinery to inform the apex court as to whether the industries ordered by it to be shifted have actually done so or not. An environmental squad under the Central Pollution Control Board had been indicated by the apex court to be set up by the Union Government. That remains to be fully effective. Meanwhile the backlash has already started and many issues are now again awaiting a decision of the apex court.

The entire judgment on aquafarming relating to the use of India's long coastal belt has been stayed and the matter is to be heard once again. The apex court's judgment concerning measures for the protection and preservation of the Taj remains to be fully implemented. In the Oleum gas leak case the Supreme Court had observed that life, public health and ecology have priority over unemployment and loss of revenue. This implies a radical change in the entire economic direction of the State. Between the pronouncement of law and the reality of the political economy lies an evolution in which the life of law may not be logic but experience.

20

Social and Environmental Accounting and Auditing

TRILOK KUMAR JAIN

INTRODUCTION TO CONCEPTS

Social and environmental accounting implies documentation and reporting of all investments, expenditures, utilization, and disposal of resources obtained from society (for social accounting) and from environment (environmental accounting). The primary purpose of such accounting is to assist management in getting familiar with the impact the organisation has made on society/environment and to assist in the process of social/environmental audit.

Audit: an examination of books, accounts, and vouchers of a business, as well enable the auditor to report whether he is satisfied that the balance sheet is properly drawn up, so as to give a true and fair view of the state of the affairs of the business, and that the profit and loss account gives a true and fair view of the profit or loss for the financial period, according to the best of the information and explanations given to him, and as shown by the books; and if not, to report in what respects he is not satisfied.[1]

Environmental Audit aims at verifications and validation to ensure that the various environmental laws are complied with and adequate care has been taken towards environmental protection and preservation.

Social Audit is verification, validation, measurement, evaluation and reporting of the organization's performance in fulfillment of its social responsibility.

Some other Related Concepts

Energy Audit

It tries to study, verify, evaluate and report the organization's efforts at conserving energy and energy utilization.

Shadow Price

Many times market prices do not reflect true value of a goods. This may be due to monopoly or regulated prices. Shadow prices are the prices which reflect the real value attached to each commodity. Shadow prices reflect the opportunity cost of the goods. These prices are also called accounting prices, social prices or planning price. Traded commodities are valued at their border prices (world prices, in free trade), non-traded goods are valued at their marginal cost of production.

Standard Conversion Factor (SCF)

Developing countries are many times not willing to adopt open market policies. In order to arrive at appropriate prices, SCF is used for calculation of appropriate prices.

Social Weights

Different outputs have different values to different segments of society, many times not measurable in money terms. For example, a benefit of Rs 100 for a poor person is of greater value than the same benefit for a rich person. Such benefits are given some weightage to arrive at their true value. Social weights are assigned in SCBA for calculation of the value of output.

Social Rate of Interest

It is used in SCBA (Social Cost Benefit Analysis) for evaluation of a project. It is the opportunity cost of society. It is calculated on the basis of present and future cash flows of alternative projects. If the amount had been invested in some other project, what would have been the future flow of funds.

NEED AND RATIONALE OF SOCIAL AND ENVIRONMENTAL AUDIT

Social and environmental audits are not yet very popular in India, however, they are in use for a long time in developed countries. The reason for their adoption lies in very active public interest groups and social activists. In India, the demand for social audit and environmental audit is increasing.

METHODS OF SOCIAL AND ENVIRONMENTAL ACCOUNTING

As discussed earlier, social and environmental accounting is the documentation and reporting of all expenditure and investment on resources of society/environment. The process of such accounting is slightly different from conventional accounting. The principles of accounting are applicable, but the objective and methods of preparing final statements are different.

Important models are as under;

- Linowes Model
- ABT Associates model
- SCBA

Linowes Model

David F. Linowes (a CPA or Certified Public Accountant) has given a model which incorporates Socio Economic Operative Statement (SEOS). The model suggests that only those expenses, which have been undertaken by the organisation voluntarily for the development of society or for environment preservation should be shown in the statement. The expenditure which have been undertaken for social development or environment protection under some law or mandatory requirements should be shown as a footnote only.

The items like cost of installing pollution control plants (voluntarily), training employees are put in social improvements.

The format of the statement is as below:

ABT Associates Model

As per this model, the balance sheet would contain list of social assets on one side and list of social liabilities and equity on the other side. The income statement will contain list of social benefits and social costs.

TABLE 1

Proforma of S.E.O.S. as given by Linowes

Socio Economic Operative Statement	*Amount*
Social-Actions-People Related	
Improvements	
Total Improvements	
Less Detriments	
People Related Actions: Net Improvement for the Year	
Social Actions—Environment Related	
Improvements	
Environment Related Actions: Net Deficit for the Year	
Social Actions—Product Related	
Improvements	
Less-Detriments	
Product Related Activities—Net Improvements for the Year	
Total Socio Economic Improvement for the Year	
Add: Net Cumulative Socio Economic Improvement at the Beginning of the Year	
Grand Total : Net Socio Economic Improvement	

TABLE 2

Proforma of Balance Sheet as given by ABT Associates

Assets	*Amount*	*Liabilities*	*Amount*
Staff Assets		Staff Liabilities	
Organizational Assets		Organizational Liabilities	
Use of Public Goods		Public Liabilities	
Financial Assets		Financial Liabilities	
Physical Assets		Society's Equity	
Total		Total	

TABLE 3

Proforma of Social and Financial Income Statement as given by ABT Proposal

Social and Financial Income Statement	*Amount*
Benefits	
To company/stockholders	
To staff	
To clients/general public	
To community	
Total	
Costs (Expenditure)	
To company/stockholders	
To staff	
To clients/general public	
To community	
Total	
Net Income	
To company/stockholders	
To staff	
To community	
Total Net Social Income	

SCBA

UNIDO has given some guidelines on Social Cost Benefit Analysis. These guidelines are also useful for environmental and social accounting. There are also some other guidelines available like one issued by OECD, by Planning Commission (1975) etc. SCBA documents social costs, social interest rate, shadow prices etc. in preparing feasibility report of any project. International organisations like World Bank etc. accept a project if its social contribution is positive. Such social contribution has to be calculated as per SCBA norms.

The methods of SCBA are as under:

- measurement of economic, social, and environmental inputs

and outputs from a project.

- calculation of social accounting or shadow prices on inputs or outputs from the project.
- selection of appropriate social rate of interest.
- computing social profitability of project.

Little and Mirrless (1977) have given a formula for calculation of Standard Conversion Factor

$$SCF = \frac{O^{*}AP}{Q^{*}MP}$$

SCF= Standard Conversion Factor

A = Available supply

AP=Accounting prices

MP=Market prices

METHODS OF SOCIAL AND ENVIRONMENTAL AUDIT

Social Audit

Following are the principal methods of social audit:

- Constitutency group attitudes audit
- Corporate rating audit
- Macro-micro social indicator audit
- Social process audit
- Partial or comprehensive audit

Constituency Group attitudes Audit

There are a number of stakeholders and pressure groups attached with any organisation. The stakeholders include customer, suppliers, shareholders, employees, creditors, associates and surrounding community. The method of constituency group attitude uses the opinion of various stakeholders in audit process. The method adopted is as following :

- Identification of priorities for constituency reference groups.

- Development of dimensions based on the criteria identified (qualitative or quantitative).
- Preparation of social profits/losses based on the dimensions identified above through discussion with executives of the organisation.
- Identification of satisfaction of representatives of social groups on various profiles.
- Analysis of preferences given by respondents and preparation of preference areas of organisation for social good.

Corporate Rating Audit

A method similar to previous method, it also relies on opinion poll substantially. This method has been initiated by social activist organisations, mutual funds, church groups and universities in America. Leading mutual funds, Universities etc. have evolved their methods of undertaking audit using detailed studies on organization's impact on environment and society and based on opinion poll.

Important corporate rating organisations in USA are as under:

- Interfaith Centre on Corporate Responsibility (ICCR)
- Council on Economic Priorities
- Yale University
- Third Century Fund
- Caux Round Table (CRT) Committee

Macro-micro Social indicator Audit

Macro social indicators are those indicators which have been set up by a particular society/nation as goals of social life and as measures of judging performance.

Micro social indicators are those indicators which as an organisation has set up for itself as goals for social/organizational development and measures of performance evaluation in terms of social good. The micro indicators are with reference to social indicators.

In the method of macro-micro social indicator audit, the organisations are evaluated on the basis of their contribution to social, development as measured in terms of social development.

This method became popular when The First National Bank of

Minneapolis (USA) used it for its audit.

Social Process Audit

In this method, all those programs are evaluated which are undertaken for social reasons. The basic premise of this method is that the community is interested in organization's involvement in social development. Therefore organisations must have some programme for social development and they must be periodically appraised.

This method became popular due to its use by Bank of America.

- Assessment of circumstances under which each social program audited came into being.
- The goals of social program being audited.
- Rationale involved, how the company plans to attain its goals.
- What is actually done as opposed to what ought to have been done.

Bauer and Fenn have also given a comprehensive management audit manual which can be used by those undertaking such audit.

Partial or Comprehensive Audit

Partial audit is focused on a small segment like energy audit, environment audit etc. Comprehensive audit on the other hand includes all segments of audit like energy audit, financial audit, management audit, social audit etc. it is most comprehensive and all pervasive audit.

Environmental Audit

Environmental audit is undertaken in the following steps:

- documents and report preparation and data collection about the compliance of environmental laws and efforts aimed at environmental preservation.
- presentation of the report to independent auditor
- verification and validation by auditor
- reporting by auditor

Documents and report preparation and data collection about the compliance of environmental laws and efforts aimed at environmental preservation

This aspect is related to social and environmental accounting. A committee may be appointed from within the organisation or an independent team of professionals may be hired for this task. If in-house committee is appointed, care must be taken in inducting persons from different disciplines and backgrounds in the committee. The committee should have accountants, sociologists, environmental law experts etc. In large organisation, the environmental/social accounting may be undertaken by their accounting department also as a routine activity. But, for this, adequate orientation and training should be given.

Presentation of the report to independent auditor

The accounts and report prepared by the committee should be presented to an independent auditor.

Verification and validation by auditor

The auditor may personally visit the plant and verify the pollution control equipment. He may also meet employees and stakeholders and verify company claims.

Reporting by auditor

The auditor presents the final report to the company.

INDIAN PERSPECTIVES

There were very few organisations undertaking environmental audits and there were a few auditors available. However, in the last decade the importance of this field is constantly increasing. This audit is now being adopted by many companies.

Rule 14 of the Environment Protection Rules 1986 (as introduced on 13.3.92) requires certain industrialists to submit environment audit report every year. The report has to be in a particular format (Form V). The report is mandatory in following cases:

- for every industry/operation/process which require consent under section 25 of Water (Prevention and Control of

Pollution) Act 1981.

- for every industry/operation/process which require consent under section 2 of Air (Prevention and Control of Pollution) Act 1981.
- for every industry/operation/process which require authorization under Hazardous Wastes (Management and Handling) Rules 1989.

The environment audit report has to be submitted to State Pollution Control Board.

Other Aspects relating to Environment Protection in India

The government has enacted following legislations/schemes to ensure environmental protection and preservation.

- Air (Prevention and Control of Pollution) Act, 1981
- Water (Prevention and Control of Pollution) Act, 1981
- Environment Protection Act, 1986
- Ecomark Scheme
- The National Environment Tribunal 1992

The government has set up Central and State Pollution Control Boards and Laboratories under the above laws. No industry can be set up without taking prior environmental clearance. The Pollution Control Boards are given wide powers to collect samples, inspect plants and prevent pollution control.

Amongst indian companies, TISCO is pioneer in conducting social audit. However, we also find examples of Bhopal Gas Tragedy. There are many industries in India which are causing massive environmental pollution (for example toxic wastes of some units is responsible for death of fish in Indian sea beach, textile, chemical and sugar units in many cities like Pali are emitting chemicals and other effluents in open spaces.) Many industries are using child labour (for example carpet industry of Varanasi, Akik Industry of Gujarat, Fireworks industry of Shiwakasi, Glass industry of Moradabad etc.)

Developed countries have imposed tight environment related legislations in their countries. Many MNCs are shifting their production base and chemical plants to developing countries in order to escape the stringent requirements. The developing countries have inadequate monitoring and regulating agencies. The general public is also not united and vocal. During April 1996—January 1997, 15000 tonnes of lead and battery waste and 12000 tonnes of zinc wastes were imported into India. These wastes are toxic in nature and must be safely buried as per the standards of developed countries. Greenpeace and other international organisations launched protests against such transactions (Indian activists were not even aware of these).[2]

As per a Worldwatch Institute report, multinational tobacco companies are facing legal problems in developed countries and hence shifting their focus on developing countries like India. The report described it as modern 'opium war'. The report predicted that for every 1000 tonnes of tobacco goods produced, 1000 persons would die eventually from tobacco related diseases.[3]

World Bank asked Indian power companies in 1997 to rehabilitate project affected people. It took this step following a complaint by an NGO. World Bank even warned power companies that fresh loans will be issued only of power companies get the clearance from the environment cell of World Bank. Following the threat of World Bank Powergrid Corporation decided to throw open its policy on environment and social issues for public vetting.[4]

Of late, we find the courts also taking a pro activist stand in cases related to environmental pollution.

Notes and References

1. From Spicer and Pegler's Practical Auditing (Allied Publishers P. Ltd. 1975) p. 3.
2. See *Economic Times*, Dt. 17 April 1997.
3. See *Economic Times*, Dt. 25.6.1997 for details.
4. See *Economic Times*, Dt. 30.6.97 for details.

References

Baur, Raymond A. and Don H. Fenn, 1992, *The Corporate Social Audit*, Russell, Sage Foundation, New York.

Goswami, Dileep, 1980, *A Handbook of Pollution Control*, Emcon Business Review, New Delhi.

Little I.M.D. and J.A. Mirrless, 1975, *Project Appraisal and Planning for Developing Countries*, Oxford, New Delhi.

Puttaswamaiah, K. (edited) 1980, *Project Evaluation: Criteria and Cost Benefit Analysis*, Oxford and IBH, New Delhi

21

Environmental Disclosure in the Corporate Annual Reports

MANGALENDU NARAYAN ROY

1. INTRODUCTION

In the evolving scenario of the globe; legal, political cultural, social and other environments keep changing in each country as time passes. As a result, the rapid growth and development of corporate organisations with technological advancement has been experienced and as such the role of modern business in society has been changing dramatically. The large companies, once looked upon as the exclusive concern of its owners, is coming to be viewed as an instrument of society. Managers are discovering that maximisation of return to share holders is not a sufficient goal; society is demanding more. In modern times, "business success is judged, not simply by a company's technological or financial performance, but by how well that company interacts with and serves social, legal, political, governmental and broad human interest" (Fredrick *et. al.*, 1948). Again, business executives who make their plans for the present and the future, determine their strategies, take their decisions and ultimately who wish to direct their companies in the successful pursuit of profits must take into consideration the information regarding the broad social environs. These changing scenarios and resultant growth and development of corporate organisation have given rise to the unprecedented need for information of various dimensions as present day demand. In other words, there has been a sustained demand for relevant and reliable information to the decision-makers and as such it is an age of information. Converting these various information into useful knowledge, business executives are taking decisions for attaining their objectives (Fredrick *et. al.*, 1988). Accounting, as an information system, is designed to play a dominant role in such decision situation.

Now accounting is in an age of rapid transition. Its environment has undergone vast changes in the last two decades. Changing social attitudes combine with development of information technology, quantitative methods and the behavioural sciences to affect radically the environment in which accounting operates today, thereby creating the need to re-evaluate the scope and objectives of accounting in a wide perspective. (Glautier and Underdown, 1984). During the 1980s an ever increasing emphasis on the pursuit of financial efficiency and value for money as the keys to achiving social prosperity led inevitably to accounting, and indeed accountants, occupying a position of prime importance in the public policy arena (Owen, 1992). As such the changing environment has extended the boundaries of accounting and it is moving away from its traditional procedural base, encompassing record-keeping and such related works as the preparation of budgets and final accounts, towards the adoption of a role which emphasizes its social importance.

In this context, academicians, accounting practitioners, research scholars, professional institutes and similar other organisations started research work on accounting and reporting seriously since early sixties and various new dimensions have been added to the accounting discipline. One of them is "accounting and reporting for conservation of environment" which becomes synonymous with 'environmental disclosure', 'environmental Pollution accounting and disclosure', 'Pollution control accounting', 'accounting and reporting for environmental Protection', 'green reporting' etc. In this article, I have made an attempt to highlight the idea regarding conservation of environment (Pollution Control etc.) and have discussed the environmental movement, environmental issues of the business and their disclosure in the annual reports as a current corporate agenda all over the world. I have also made a case study on environmental disclosure by the Indian companies in their annual reports.

II. ENVIRONMENTAL MOVEMENT AND BUSINESS

Nature by itself, the Industrial Revolution, the higher standard of living, population explosion, changes in social values etc. have created an environmental and ecological crisis. As a contributor to pollution, business is also involved in society's ecological problems. In the obsession for profit maximisation, industrial units have played havoc with environment and as such "all industrial societies whether the United States, Japan, the Soviet Union or France—create a portion of the world pollution and waste, by polluting air, contaminating water, damaging the soil etc., simply because these are the unavoidable by-products of a high level of industrial activity. (Rifkin 1980). This has created unprecedented health problem for the people and caused enormous damage to human and animal habitats.

"It is interesting to note that the genesis of modern environment movement (green movement) can be traced back in 1962 to the publication of a book, 'Silent Spring' by Rachel Carson (1962)" (Bose, T.K., 1995). Ms. Carson, a biologist, wrote forcefully of her scientific alarm about the widespread, reckless use of pesticides, particularly DDT. She described in her book how the widespread use of insecticide DDT led to the disappearance of many songbirds in America. Continued heavy use of such chemicals, she believed, posed a direct, long term threat not just to the target insects but to all animals that belonged to interwoven food chains in nature; eventually, the effects would be felt by humans themselves. After a decade of sustained movement the use of DDT was banned in America in 1972 (Bose, T.K., 1995).

Conservation of environment (Pollution Control etc.) became a high social priorities in the United States during 1970s. In the 1980s and 1990s, environmental movement and action programmes have continued to receive strong public support all over the world. In India, as elsewhere business firms are coming under tremendous pressure from environmental activists, social workers, judiciary and the government to conform to the strict environmental laws and regulations or face dire consequences which include penalties, imprisonment and closure of business. Mr. M.C. Mehta, Mr. Sunderlal Bahuguna, Ms. Medha Patker and Subhas Dutta among other, are waging relentless battle against environment pollution creating greater public awareness regarding environmental issues (green issues) and forcing the Central and State governments to take action against polluting units.

As a matter of fact, a fierce movement for the protection of environment has been gathering momentum all over the world in the wake of many adverse consequences resulting from the pollution. Although the general public continues to favour environmental protection, there is a growing awareness that competition, productivity and jobs also must be considered. The social challenge to business in an industrial society today, therefore, is not to stop all pollution and waste but to reduce its volume, lessen its burden on society, help to clean it up once it occurs and ultimately to find a happy medium between industrial production and nature's limits (Fredrick *et. al.*, 1988).

The environmental movement is addressing issues vital to our quality of life on earth. The momentum of such movement is creating a permanent shift in the way business operates and there is evidence of increasing co-operation between business and environmentalists as business (society's major economic institution for production of goods and services) has to line in harmony with nature, thereby cannot ignore the natural environment. So it is rightly observed that, "in the nineties environmentalism is the cutting edge of social reform and absolutely the most important issue for business" (Sullivan, 1992). But in this situation society needs to be cautious that it does not become overzealous and excessive in its demands for virtually zero ecological effects from business activities. Almost any human activity has an ecological impact. The idea is not to prevent activity, but rather to assure that activity is in harmony with nature so that nature's powers of self-restoration will maintain a clean and livable environment. However, the problems on the issue as such are complex, their solution requires many trade-offs between economic production and a cleaner, safer environment and the task for environmental protection will continue to be a challenge for business in the comming 21st Century.

III. ENVIRONMENTAL LAWS AND EMERGING CORPORATE RESPONSIBILITY IN INDIA

In the wake of growing national and international pressure for preventing and controlling environment pollution in India vis-a-vis conservation of environment, a number of new acts have been passed and old ones have been amended to ensure strict compliance with environmental norms. The Environment Protection Act, 1986, the Public Liability Insurance Act, 1991, have been pased and the Water

(Prevention and Control of Pollution) Act, 1974, the Air (Prevention and Control of Pollution) Act, 1981. The Wildlife (Protection) Act, 1972, The Motor Vehicle Act 1938, The Forest (Conservation) Act, 1980 have been amended. The Ministry of Environment and Forests has announced a 'Policy Statement for Abatement of Pollution' in February 1992 according to which, the key elements for pollution prevention are adoption of best available clean and practicable technologies rather then end of pipe treatment. In addition many action plans have been formulated covering a wide ambit of subjects such a clean technologies, improvement of water quality, institutional and human resource development, forestry and natural resource accounting. Many monitoring agencies like Central and State Pollution Control Boards have been set up for controlling pollution. "An Eco-Mark" label has been introduced to lebel consumer products that are environment friendly. A notification making environmental audit mandatory has been issued which requires all industries applying for environmental clearance to submit an annual environmental audit report to the concerned State Pollution Control Board. Companies Act, 1956 is proposed to be amended to include an 'Environment Statement' in the Annual Report of Companies (India, 1993). A comprehensive Bill to set up Environmental Tribunals is pending before the Parliament. Recently (16.04.96), Honourable Justice Kuldip Singh of Supreme Court in his judgement in the context of Howrah Pollution affairs, has directed the Calcutta High Court to form 'Green Bench' to resolve the issue (Biswas, 17.06.96).

Both environmentalists and public opinion have become major forces in shaping environmental legislation that have a significant impact on business units and in the field of research work in accounting. The implication of the current environmental laws and regulations is that business firms will have to take the initiative and bear the burden of controlling the pollution and conservation of environment for which they are responsible. Firms are now classified into 'green firms' and 'brown firms' on the basis of their response to comply with the environmental regulation. Firms which are sensetive and responsive to the environmental norms and offer to adopt measures for controlling pollution on voluntary basis are called 'green firms'. Whereas the firms which are unwilling to comply with the pollution norms and seek to escape responsibility in this regard are known as 'brown firms'. The classification is important for the monitoring agencies which verify the performance of companies with regard to the environmental requirements. Therefore, the environmental responsibility is no longer an optional responsibility to the business.

The current concern for 'Conservation of environment' (Pollution Control etc.) as a mandatory obligation for an enterprise has created a fresh urgency for the accounting profession to devise effective ways and means of measuring and reporting the costs of conservation of environment so that the management can steer clear of impending threat to business and contribute as a responsible citizen to the preservation of the fragile eco-system of our planet.

IV. ENVIRONMENTAL ISSUES AND THEIR DISCLOSURE IN THE CORPORATE ANNUAL REPORTS—AN OVERVIEW

Environmental issues (conservation of environment etc.) being an important area of corporate social responsibility had been widely advocated in literature continuously by Carson (1962), Indian International Centre (1966), Bowman (1975), Davis (1976), Beesley *et. al.* (1978), Fred Luthens (1980), Davis (1983), Goodin (1983), Mathews (1987), Ryle (1988), Archibugi (1989), Katyal (1989), Kirkpatrick (1990), Weisner (1990), Institute of Business Ethics (1990), Dan (1991), Plant *et. al.* (1991), Hansell (1991), Carson *et. al.* (1991), Davis (1991), Sullivan (1992), Wicks (1992), Hopfenbeck (1993), Shrivastava (1993), among others.

Environmental disclosures in the corporate annual report assumes that the organisations are socially conscious and are ready to discharge such social obligations for the well being of the society. Numerous proposals had been made over years by accountants and researchers to develop methods to measure and report on conservation of environment [Beam and Fertig (1971), Abt (1971, 1972, 1974, 1975), Marlin (1973), Brummet (1973), Linowes (1973), Nikolai (1976), Dierkes *et. al* (1977), Burke (1980), Preston (1981), Kelly (1981), Owen (1981), Denuyl (1984), Guthrie and Mathews (1985), Gray *et. al.* (1987), Sinha (1988), Jaggi (1989), Lickiss (1991), Owen (1992), Spellerberg (1992), Ing (1992).

Several attempts had also been made from 1970s, to develop a normative theory of accounting and reporting for conservation of environment as an area of social reporting [viz. Churchil (1974), Seidler and Seidler (1975), Estes (1976), Ramanathan (1976), CICA (1978), Johnson (1979), Tinker and Lowe (1980), DeFillippi (1982), Lal,

Jawahar (1984), MacIntosh (1985), Gambling (1985), Gray *et. al.* (1986), Owen *et. al.* (1987), Turner (1988), Guthrie and Parker (1989), Gray (1990), Owen (1990), Someya (1991), Chattopadhyay (1991), Bernsterin (1992), Owen (1992), Gray (1992), Owen (1993)]. Efforts are also being made in some universities to develop framework for analysis of corporate social responsibility accounting and reporting (where environment is considered as an area) [Porwal-1993].

Further, the form and content of corporate disclosures on conservation of environment and its measurements had been the subject of series of public hearings and pronouncements by the Security Exchange Commission (SEC-USA) and some major Accounting Association in different countries of the world. In the USA, from 1971 to 1973, the SEC had issued a number of releases (SEC 1971, 1973) concerning environmental disclosures. It involved a great deal of research work in this direction. The American Accounting Association (AAA) Committee on Environmental Effects of Organisation Behaviour (AAA-1973) had proposed narrative disclosure, in footnotes to the audited financial statement of information concerning environmental issues. The AAA Committee called for the verbal description of—

(a) identification of environmental problem,
(b) abatement goals of the Organisation,
(c) progress of the organisation towards attainment of environmental goals,
(d) disclosure of material environmental effects on financial position, earnings and business activities of the organisation.

Another Committee was formed by the AAA (AAA-1974) for measurement of social cost to the business where environment is one of the areas of social activities. The National Association of Accountants (NAA-1974) committee on accounting for corporate social performance (1974) in its efforts to resolve the issue, had proposed four major areas of social performance, one of which was 'physical resources and environmental contribution'. Activities directed towards preservation of environment, conversation of scarce resource, disposal of solid wastes etc. are considered in this area. In United Kingdom (UK) the 'Corporate Report' (1975), published by the Accounting Standard Steering Committee (ASSC-1975) of the Institute of Chartered Accountants in England and Walses (ICAEW) and the 1977 green paper 'the Future of Company Report' (HMSOCMND 6888-1977) had drawn attention to the concept of social—responsibility reporting (where environment protection is an issue) and made an attempt to move in that direction. In Canada, the Stamp Report (1980) published by the Canadian Institute of Chartered Accountants (CICA-1980) had discussed the broader accountability concept where the ranges of users were also broader. Efforts are also made by the accounting bodies of some other countries in varying degrees.

The increase in social pressure for conservation of environment in the 1960s, environmental disclosures have become gradually popular throughout the world. In spite of difficulties in measuring and reporting environmental effectives, a number of companies have been trying in their own way, to develope suitable methods for meaningfully informing people more about what they have done so far on the issue vis-a-vis their social responsibilities.

Most of the work in this respect has been done in the USA. Various renowned companies of USA lilke Scovill Manufacturing Company. Atlantic Richfield Company, Jones Corporation, Eastern Gas and Fuel Association, U.S. Steel, etç. have published report over years on conservation of environment as a part of their social activities under different titles and in different manners. It appears primarily in the annual reports, but a significant number of companies use separate environmental action report or separate social report.

A number of companies in United Kingdom (UK), like The Body Shop Internation PLC, Caird Group PLC, Argyll Group PLC, British Gas PLC Halma PLC, Tesco, PLC, North West Water Authority etc. have published report on conservation of environment under different titles such as Environment Audit Report, Pollution Control Statistics, Water Quality Report, Genuine Commitments, Good Corporate Citizenship Social Report etc. and still they are doing so (Harte and Owen). Such reports are being regularly reviewed in social audit Magazine which is published by the Public Interest Research Centre. Some companies in UK like VOLVO, South West Water Co. Hewlett Packard, ICI, Procter and Gamble etc. have started publishing Environmental Policy Statement in their annual reports (Blaza, 1992).

Developments are also taking place in the Federal Republic of Germany, Netherlands, Sweden, Spain, Japan and some other countries. The Australian Companies are disclosing social responsibility information in the areas of environment, energy, human resources, products, community involvement so the idea of conservation of environment is gaining ground fast all over the world. A piece of legislation has been framed in connection with social accounts in France, where conservation of environment is considered as an area. But the only European country with any specific disclosure requirements in the area of environmental impact information is Norway. The Enterprise Act was amended in 1989 to require corporations to include, in the board of directors' report, information on emission levels, Contamination and details of measures both planned and actually performed by the corporation with the objective of cleaning up the environment.

From the above discussion, it is evident that the environmental issues and their disclosure in the corporate annual reports are considered by the academicians, researchers, professional managers, accountants and Professional bodies as a basic common agenda of the business from about mid of this century. Discussion on the issue from various angles have been started from 1960s and they are still going on. The demand for better deal from the business community is also slowly building up and the concept of conservation of environment is being accepted by some of the enlightened business corporation. This has led to a state where companies are trying to prepare report on conservation of environment separately or trying to add a dimension to their financial reporting in this direction and this is gaining popularity amongst the companies around the globe. This increasing trend in the number of report on conservation of environment, however, is not indicative of the quality of such report as most of the companies produce the same in a descriptive manner.

V. ENVIRONMENTAL DISCLOSURE IN THE CORPORATE ANNUAL REPORTS IN INDIA—A CASE STUDY

India is one of the leading developing countries in the world. Here also, increasing public awareness regarading 'Conservation of Environment' is compelling a modern corporation to redefine its functions. In tune with this current thinking, though no commonly used normative approach has been developed in India, no government directives or legal compulsion, there has been a humble beginning of reporting on social issues in general and more particularly that on environmental information by some progressive companies in India both in Public and private sector to provide their information on conservation of environment in published annual reports and/or through separate means of disclosure.

Since the 1970s, a number of studies have been carried out concerning the corporate social disclosure in general to examine the levels and types of social information provided in the annual reports. Maxwell and Mason (1970), Lesson (1977), Rey and Dierkes (1978), Trotman (1981), Brooks (1986) Porwal and Shrama (1991), Roberts (1992) etc. have all examined the extent to which companies produce social information of which environmental information is a part. However, specific studies relating to environmental disclosure in the annual reports are not many. Prominent, among the studies conducted in India and abroad, are of Norman Pope (1971), Williams *et. al.* (1972), Spicer (19780, Ingram and Frazier (1980), Sengupta (1988), Roberts (1992), Harte Lewis and Owen (1991), Eresi (1996) etc. These have ranged from surveys of frequencies and types of pollution disclosures to discursive case studies of such practice. My study also aims to examine the Indian practice in respect of environmental disclosure in the corporate annual reports. For this purpose, I have selected 55 companies, of these companies, 25 are in public sector and 30 are in the private sector representing manufacturing, trading, construction and servicing industries.

Research Methodology

The sample for the public sector companies has been selected from the Index of companies published in the Bureau of Public Enterprise Report—1990-91, Govt. of India. The total list consists of 236 public sector companies engaged in manufacutring, trading, construction and servicing are treated as the population. From it, a random sample of 25 companies has been taken for study. The sample for the private sector companies has been selected from the 'Index of companies' published in the Calcutta Stock Exchange Year Book 1990-91. The total list consists of 221 companies. Out of which 3 companies engaged in financial service have been shortlisted. This has been done to exclude companies which are basically being treated as investors (institutional investors). After shortlisting, 218 private sector companies engaged in manufacturing, trading, construction and servicing are treated as the population. From it, a random sample of 30 companies has been taken for study.

A checklist of six items relating to environmentally responsible activities has been drawn up to determine the disclosure levels of companies surveyed. The checklist used has been adapted from McComb 1978 survey (McComb, 1978) with some modification to suit Indian condition. The items of the checklist are : (i) Pollution Control, (ii) Recycling and reusing disposable materials, (iii) Conservation of energy, (iv) Development of new energy resources, (v) Preservation of ecology of forest tree plantation, (vi) Other environmental action support programmes. The annual reports of those sample companies for three years ending in 1987, 1991, and 1996 have been then scrutinised. Items of environmental concern disclosed in the report are noted on the checklist. The results have been analysed in two ways: first, the number and percentage of companies which have disclosed data on conservation of environment in different places of annual reports are recorded, second the number and percentage of companies which have disclosed each item of environment data in each of the years is established.

Limitation of the Study

The study has some limitations of its own. First, the study is made on the basis of very small sample of both public sector and private sector companies. A large sample covering both large and small companies may provide more dependable results. Second, the study is based on the assumption that the companies concerned about environmental responsibilities also make their disclosure on the issue. This may not always be the case. Third, the information presented by the companies was unaudited and non-specific. No judgement could be made as to the extent to which the companies presented a complete picture of their corporate activities. As a consequence, the percentage given may be a generous estimate of the true extent of corporate environmental disclosure. Four, the study does not include bank other financial institutions. The generalisation is, therefore, limited to manufacturing, trading, constrution and servicing concerns only. Five, the sample survey is biased in favour of large corporations with vast resources at their disposal with which to experiment with non-essential informations. Six, the mention of a companies involvement with community or its attempt to alleviate pollution is taken for a disclosure. Environmental issues as disclosed by the sample companies are in years 1987, 1991 and 1996 only covering a period of ten years. It is quite possible that the sample companies might have come out with greater details in mid term years.

Nevertheless, the results of the study based on this sample should give an approximation of the extent of environmental disclosures in annual reports and accounts in three different point of time (i.e. in 1987, 1991 and 1996) and of the underlying trends of reporting practices in this respect over years amongst the Indian largest Companies.

Analysis of Survey Results

An analysis of the annual reports of the sample companies reveals the following:

The common practice followed by the Indian companies of both the sectors over years regarding environmental disclosure is to offer descriptive information and itemise the same in the Directors' Report. A few companies disclose their information through Notes and Schedules and Separate Social Account. Table 1 shows the number of companies under study of both public and private sector making environmental disclosure in different places of the annual reports at the three point of time.

Table 1 indicates that the trend of number of companies making environmental disclosures through Directors' Report are increasing

TABLE 1

Table showing the Number of Companies under Study Making Environmental Disclosure in Different Places of Annual Reports

Disclosure Practice	*Public Sector Cos*			*Private Sector Cos*		
	1987	*1991*	*1996*	*1987*	*1991*	*1996*
Total Number of companies under study	25	25	25	30	30	30
Location of Environmental Disclosure	Number of companies making environmental disclosures					
In Directors* Report	10 (40)	21 (84)	25 (100)	13 (43)	30 (100)	30 (100)
In Notes and Sch.	5 (20)	5 (20)	5 (20)	2 (6.67)	NIL	NIL
In Separate Section (Social accounts)	2 (8)	2 (8)	2 (8)	NIL	NIL	NIL

Source: Published Annual Reports of 55 Companies for the periods ending in 1987, 1991, and 1996 (Results Computed—Percentage of companies making environmental disclosures are shown in the brackets).

over years in case of both the sectors. In private sector, all the companies under study are making some short of environmental disclosures through Directors' Report in 1991 and 1996 and that in case of public sector only in year 1996. Only 5 companies are making disclosure through Notes/Schedules at each of the three point of study in case of public sector companies. In case of private sector companies only two companies in 1987 are making such disclosure through Notes/ Schedules. Disclosures through separate section (social accounts) of annual reports are made by two companies (ONGC and SAIL) in 1987 and by two companies (SAIL and BHEL) both in 1991 and 1996 amongst the public sector companies understudy. No such account was prepared by any private sector companies understudy over the period. Companies of both the sectors disclosing information through Notes/Schedules and separate section over years are making disclosures in quantitative or monetary terms. The study shows that disclosures on narrative terms are significantly more than that disclosures on monetary terms.

The itemwise disclosure on the basis of checklist made by the companies of the both sectors under study in their annual reports relating to 'Environmental Disclosure' over years is shown in Table 2.

Table 2 exhibits that in the public sector, the number of companies making disclosures under item nos. 1, 2, 3, 5 and 6 has been increasing over years where as in the private sector, the number of companies increasing over years. The number of companies making disclosure under item No. 4 is equal in 1991 and 1996 in the public sector, whereas in the private sector, the number of companies making environmental disclosures under item nos. 3, 4 and 6 is equal in 1991 and 1996. Information in relation to 'recycling and reusing disposable making environmental disclosure under item no. 1 and 5 has been materials' (item no. 2) has been disclosed by none of the companies in the public sector in 1987 and that in the private sector in 1991. Another thing has been revealed from the analysis that the percentage of companies, disclosing environmental information barring a few remains consistently low in both the sectors. However, number of public sector companies disclosing environmental responsible activities are more than that of private sector Companies.

TABLE 2

Table showing Analysis of Item-wise Disclosure Pattern in the Annual Reports of the Companies Under Study

Disclosures on Environments	*Number of Companies making Disclosures*					
	Public Sector			*Private Sector*		
	1987	*1991*	*1996*	*1987*	*1991*	*1996*
Total number of Cos.	25	25	25	30	30	30
1. Pollution Control	6	12	16	8	5	9
	(24)	(48)	(64)	(26.64)	(16.65)	(30)
2. Recycling and reusing	—	1	4	1	—	1
disposable materials	(4)	(16)		(3.33)		(3.33)
3. Conservation of Energy	5	20	25	1	30	30
(legal requirement after 1988)	(20)	(80)	(100)	(3.33)	(100)	(100)
4. Development of new	1	2	2	1	2	2
energy resources	(4)	(8)	(8)	(3.33)	(6.67)	(6.67)
5. Preservation of ecology	8	10	18	6	3	10
of forest/tree plantation	(32)	(40)	(72)	(20)	(10)	(33.33)
6. Other environmental action	4	6	7	4	6	6
support programmes	(16)	(24)	(28)	(13.32)	(20)	(20)

Source: Published Annual Reports of 55 Companies under study (Result Computed: Percentage of number of Companies are shown in the bracket).

Table 3 indicates that all the companies under study in both the sectors are making some sort of environmental disclosure either in Directors' Report and/or Notes and Schedules in the year 1991 and 1996 which are substantially low in 1987 both in Public Sector (56%) and Private Sector (43%) companies. However, close study of the annual reports in respect of environmental disclosure reveals the following:

(a) Disclosures on environmental activities made by the companies of both the sectors under study are very descriptive.

(b) Though some of the companies under study have disclosed quantitative information on the issue, neither any item-wise breakup of expenditure nor its accounting treatment is shown in any of the reports.

(c) Almost all the companies of both the sectors have disclosed in full qualitative and financial information in respect of Energy

TABLE 3

Environmental Disclosures

	Public Sector Co.s			*Private Sector Co.s*		
	1987	*1991*	*1996*	*1987*	*1991*	*1996*
No. of Co.s making environmental disclosures	14 (56)	25 (100)	25 (100)	13 (43)	30 (100)	30 (100)
No. of Co.s not making environmental disclosures	11 (44)	NIL	NIL	17 (57)	NIL	NIL
	25	25	25	30	30	30

(25 public sector companies and 30 private sector companies under study. 1987—1996).

Source: Published Annual Reports of 55 Companies for the periods ending in 1987, 1991 and 1996 (Percentage of companies making and not making environmental disclosures are shown in the bracket).

Conservation in 1991 and 1996, since it is statutorily required under section 217(1)(e) of the Companies Act, 1956, read with the Companies (disclosure of particulars in the report of Board of Directors) Rules, 1988. The extent of disclosure ranged from half page to two pages in annual reports. Further, it is also found these companies are more explicit and particular in sharing the energy information.

(d) On other environmental issues (except energy conservation) the extent of disclosure does not exceed even one-fourth page. In many cases the information was shared in just one or two sentences. The extent of disclosure is by all means tardy and is of low level reflecting poor quality.

The purpose of the study has been to ascertain how the Indian companies are responding to the emerging green agenda in terms of providing relevant information within the annual reports and accounts. But it is found, that environmental disclosure in Indian companies annual reports has not gained much importance. The minimum effort by some of the companies are not fully useful to the users of such information.

VI. CONCLUSION AND RECOMMENDATION

Conservation of environment is an expensive proposition to

business units. So the majority of business firms are reluctant to voluntarily accept the responsibility of conservation of environment and commit expenditure towards that end unless forced to do so by the law. Since national and international laws are being tightened to provide for stricter penalties for infringement of environmental laws, it would be in their interest to invest in pollution control equipment and processes. In the light of the new burden of costs to be incurred by business firms under the mandatory environmental laws and Court Judgements accounting has a bigger role to play in measuring and reporting such cost of conservation of environment. But as there is no generally accepted accounting principles and standards in this respect, the corporate disclosure practice on conservation of environment is in an embryonic stage. For improvement of such disclosure practice of both private and public sector undertaking in India and to maintain a fair degree of comparability, regarding environmental reporting, between two or more companies in a year or of a company over years, the following recommendations may be made.

1. All companies should prepare their own 'environment conservation programme statement' each year as per their requirement showing the items of environment conservation to be covered, such as, pollution control, recyling of wastes, tree plantation etc. Budgeted expenditure both capital and revenue and source of fund to be utilised should be shown in the statement. This would help to make proper control over the implementation of such programme.
2. The company should incorporate all significant information relating to conservation of environment in a separate section of the annual report, which will help the users of annual report to understand the situation clearly.
3. All aspects of conservation of environment are not quantifiable at present due to paucity of proper measurement techniques and therefore both narrative and quantitative information should be provided. Moreover, social information provided through narrative form may serve the users of general category in better way at this embryonic stage.
4. Report on conservation of environment should be duly

audited by the auditor. Though there is a problem in exact computation of the environmental impact, the audit procedure should start as compliance procedure and will slowly evolve into a substantive precedure. The external auditors can be entrusted with the job and a proper in-house environmental audit division may be carved out in the mean time.

5. Business firms are confronted not only with the responsibility of acquiring pollution control equipment but also with the problem of how to treat such expenses in the accounts and how adequate disclosure can be made. As of now, no accounting standards have been developed in this respect. As a result there exists a wide variation in the accounting and reporting practices as to the treatment of pollution control costs and expenses. Keeping in mind the importance of the disclosure on conservation of environment and the experiments carried on by the companies in India in this direction, 'Accounting standards' for this purpose should be formulated by the professional bodies and these should be mandatory.
6. Specific legal provisions should be incorporated in the Companies Act regarding the disclosure on conservation of environment to serve the purpose of reporting in a better way. A clause on the environment and ecological protection can be inserted in the Companies Act 1956 in section 217(1) after conservation of Energy etc. This in the long run will ensure a healthy living of the individuals, and incidentally it will be compelling the companies in fulfilling their obligation regarding conservation of environment as well.
7. For monitoring conservation of environment by the companies, one effective step may be to set up a separate committee (comprised accountants, economist, environmentalist etc.) entrusted with specific responsibility of managing such programmes.

In sum, we clearly have a long way to go before we arrive at a recognisable system of formal accounting and reporting for environmental impact. We can, however, expect future moves in this direction as a result of supranational influences, the actions of influential user and pressure groups and perhaps most fundamentally, simple corporate self interest. The profound importance of the entire affairs demands it.

References

ABT, Clark C., *The Social Audit for Management*, New York, Amacon, 1977.

Accounting Standard Steering Committee (ASSC) of the Institute of Chartered Accountants of England and Wales, *The Corporate Report*, London, 1975.

American Accounting Association, (AAA) *Committee on Environmental Effects of Organisation Behaviour*, The Accounting, Review, Supplement 1973, p. 110.

American Accounting Association (AAA), '*Report of the Committee on Measurement of Social Costs*', The Accounting Review Supplement, 1974, Vol. No. 49.

Archibugi, F. and Nijkamp (Eds), *Economy and Ecology Towards Sustainable Development*. Dordrecht: Kluwer Academic, 1989.

Beams and Fretig, "*Pollution Control through Social Cost Conversion*", The Journal of Accountancy, November 1971, pp. 37-42.

Beesley, Michael and Evans. Tom—*Corporate Social Responsibility*, Croom Helm, London-1978.

Bernsterin, D., *Company of Green Corporate Communications for the New Environment*, Greenleaf Publishing Winter Leaf Productions, Sidney Street, Shefield, 1992.

Biswas, Ajoy, *Article published in the Anandabazar Patrika*, 17th June 1996, p. 1 and 9.

Blaza, Andrew J., 'Environment Management in Practice' in *Green Reporting: Accountancy and the Challenge of the Nineties*, edited by Dave Owen, Chapman and Hall, London, 1990. pp. 201-214.

Bose, T.K., *Role of Accounting in Controlling Environmental Pollution*, A lecture delivered in Fourth Refresher Course in Commerce on Accounting and Finance, Burdwan University, West Bengal, July 4-24, 1995, p. 2.

Bowman, E.H. & Haire M., "A *Strategic Posture Towards Corporate Social Responsibility*" California Management Review, Winter, 1975, pp. 49-58.

Brooks, L.J., '*Canadian Corporate Social Performance*,' Society of Accountants of Canada, Taranto, 1986.

Brummet, R.L., "*Total Performance Measurement*", Management Accounting, NAA, November, 1973.

Burke, Richard C., "*Disclosure of Social Accounting Information*", Cost and Management, Vol. 54, No. 3, May/June 1980.

Canadian Institute of Chartered Accountants, "*The Why, When and How of Social Responsibility Accounting*". The Canadian Institute of Chartered Accountants, Toronto, Canada, 1978.

Canadian Institute of Chartered Accountants, "*Corporate Reporting—Its Future Evolution*" (The Stamp Report-1980), CICA, Canada, 1980.

Carson Potrick and Julia Moulden, *Green is Gold*, Toronto, Harper Business 1991.

Carson Rachel, *Silent Spring*, Boston: Houghton Mifflin, 1962.

Chattopadhyay, P., "Social Responsibility Accounting: Gaps in the Contemporary Approches" in Banerjee Bhabotosh (Ed.) *Contemporary issues in Accounting*

Research, Indian Accounting Association Research Foundation, Calcutta 1991.

Churchil, N.C., '*Towards a Theory of Corporate Social Accounting*'. Thc Accounting Review, July, pp. 516-528.

Dani, C. Boroughs, '*Cleaning up the Environment*' U.S. News and World Report, vol. 110, No. 11, March 25, 1991, p. 45.

Davis, J., *Greening Business: Managing for Sustainable Development,* Oxford Basil, Blackwell, 1991.

Davis, K., '*Social Responsibility is Inevitable*', California Management Review, Fall, 1976, pp. 14-20.

Davis, K., '*The Case for and Against Business Assumption of Social Responsibility.*' Academy of Management Journal, 1983 pp. 312-322.

DeFillippi, Robert J., Conceptual Frameworks and Strategies for Corporate Social Involvement Research, in *Research in Corporate Social Performance and Policy,* Vol. 1, pp. 1-25, edited by Lee. E. Preston, JAI press INC, London, 1982.

Denuyl, D.J., '*The New Crusaders: The Corporate Social Responsibility Debate*'. Studies in Social Philosophy and Policy No. 5 (Social Philosophy and Policy Centre, Bowling Green State University, Ohio, 1984).

Dierkes, M., and L.E. Preston, "*Corporate Social Accounting Reporting for the Physical Environment: A Critical Review and Implementation Proposal*". Accounting, Organizations and Society, Vol. 2 No. 1 1977, pp. 3-22.

Eresi, K., "*Information Disclosure in Annual Reports.*" The Chartered Accountants, January 1996, pp. 45-48.

Estes, R., *Corporate Social Accounting*, John Wiley and Sons, New York, 1976, pp. 91-107.

Foglar, H.R. and F. Nutt, "*A Note on Social Responsibility and Stock Valuation*" Academy of Management Journal, March 1975.

Frederick, Davis, Post, *Business and Society—Corporate Strategy, Public Policy, Ethics*, 6th Edition, McGraw-Hill International, New York, 1988, pp. 5, 8, 15.

Gambling, T. "*The Accountant's Guide to the Galaxy, including the profession at the end of the Universe*", Accounting Organisation and Society, 1985, pp. 415-25.

Glautier and Underdown, *Accounting Theory and Practice*, Pitman Publishing, London, 1982.

Goodin, R.C., 'Ethical Principles for Environmental Protection' in *Environmental Philosophy: A Collection of Readings* (eds. R. Elliot and—A Gare), Milton Keynes: Open University Press, 1983.

Gray, Rob, Owen David and Maunders Keith, '*Corporate Social Reporting: The Way Forward*', Accountancy, December 1986.

Gray, Owen and Maunders, *Corporate Social Reportings: Accounting and Accuntability*, Prentice Hall, London, 1987.

Gray, R.H., *The Greening of Accountancy: The Protection after Pearce,* Acca., London, 1990.

Gray, R.H., "*Accounting and Environmentalism: An Exploration of the Challange of Gently Accounting for Accountability, Transparency and Sustainability*", Accounting, Organizations and Society, Vol. 17, No. 5, 1992, pp. 399-425.

Guthrie J., and M.R. Mathews, 'Corporate Social Accounting in Australia', in L. Preston (ed.), *Research in Corporate Social Performance and Policy* (JAI Press) 1985, pp. 251-277.

Guthrie, James and Parker, Lee D., '*Corporate Social Reporting: A Rebuttal of Legitimacy Theory*', Accounting and Business Research, Vol. 19, No. 78, 1989, pp. 343-352.

Hansell, Saul., "*It's Not Easy Being Green*, Institutional Investor, January 1991, pp. 101-106.

Harte, Lewis and Owen, '*Ethical Investment and the Corporate Reporting Function*,' Critical Perspectives on Accounting, Vo. 2 No 3, 1991.

Harte, G. and Owen D., 'Current Trends in the Reporting of Green Issues in the Annual Reports of United Kingdom Companies' in *Green Reporting: Accountancy and the Challenge of the Nineties,* edited by Dave Owen, Chapman and Hall, London, 1992, pp. 166-200.

HMSO Cmmd 6888 (July-1977), *The Future of Company Report*, London.

Hopfenbeck, W., *The Green Management Revolution: Lessons in Environmental Excellence*, Hertfordshire Prentice Hall International, U.K. Ltd. 1993.

India, 1993, *A Publication of Ministry of Information and Broadcasting*, Research and Reference Division, Govt. of India, January, 1994, p. 181.

Indian International Centre, New Delhi—*Social Responsibility of Business*, Manaktalas, Bombay, 1966.

Ing, Brian, "Developing Green Reporting Systems-Some Practical Implications" in *Green Reporting: Accountancy and the Challenge of the Nineties*, edited by Dave Owen, Chapman and Hall, London, 1992.

Ingram, R. and Frazier, K., '*Environmental Performances and Corporate Disclosure*', Journal of Accounting Research, Autumn 1980 pp. 614-622.

Institute of Business Ethics, *Ethics, Environment and the Company: A Guide to Effective Action*, IBE, 1990, London.

Jaggi, Bikki "*Corporate Social Performance—Objectives, Measurement and Reporting Problems*", Indian Journal of Accounting, Vol. XIX, December, 1989, Part II, pp. 8-22.

Johnson, Harold L., *Disclosure of Corporate Social Performance, Survey, Evaluation and Prospects,* Praeger Publishers, New York 1979.

Katyal, T. and Satake M., *Environmental Pollution*, Anmol Publications, New Delhi, 1989.

Kelly. G.J., "*Australian Social Responsibility Disclosure: Some Insights into Contemporary Measurement*" Accounting and Finance, 1987, pp. 97-107.

Kirkpatrick, David., "*Environmentalism: The new crusade*". Fortune, 12, February 1990 pp. 44-52.

Lal, Jawhar (Dr.) "*Framework for Corporate Social Reporting*". The Chartered Accountant, February, 1984.

Lickiss, M., "*Measuring up to the environment challenge*", Accountancy, January 1991, p. 6.

Linowes, David F., "*Let's go on with the social Audit; A Specific Proposal*", Business and Society Review, Winter 1972-73.

Luthens Fred, *Social Issues in Business*, McMillan Publishing Co. Ltd., 1980.

MacIntosh, Norman B., *The Social Software of Accounting and Information System,*

John Wiley and Sons Inc., New York 1985.

Marlin, J.T., "*Accounting for Pollution*", The Journal of Accountancy, February 1973, pp. 41-46.

Mathews, M.R., "*The Importance of Social Accounting in any complete Accounting Education System*", in the Proceedings of the Sixth International Conference on Accounting Education, edited by Dr. K. Someya, Japan Accounting Association, Kyoto, Japan, 1987, pp. 402-416.

McComb, D, '*Some Guidelines on Social Reporting in the US*', Accountancy April, 1978, pp. 50-52.

National Association of Accountants (NAA), Report of the Committee on "*Accounting for Corporate Social Performance*", Management Accounting, Ferbuary 1974, pp. 39-41.

Nikolai, Loren A., John D. Bazley, and R.L. Brummet, "*The measure of Corporate Environmental Activity*", National Association of Accountants, New York, 1976, Summary appeared in Management Accounting, June 1976, pp. 38-40.

Owen, D., "*Why Accountants can't Afford to Turn their backs on Social Reporting*", Accountancy Vol. 92, No. 1049, January, 1981, pp. 44-45.

———, "*Towards a Theory of Social Investment: A Review Essay*". Accounting Organizations and Society, Vol. 15, No. 3. 1990.

———, "*Green Reporting: Accountancy and the Challenge of the Nineties*", Chapman and Hall, London 1992, p. 3.

———, "The Implication of Current Trends in Green Awareness for Accounting Function: An Introductory Analysis" in *Green Reporting: Accountancy and the Challenge of the Nineties*, edited by Dave Owen, Champman and Hall, London, 1992, pp. 3-30.

———, "*The Emerging Green Agenda: A Role for Accounting" in Business and the Environment: Implications of the New Environmentalism,* (ed. D. Smith) Paul Chapman Publishing, London, 1993.

Owen, D., Gray, R. and Moundress, K., "*Researching The information Content of Social Responsibility Disclosure: A comment*", British Accounting Review, August 1987, pp. 169-75.

Plant, Christopher and Plant Judith (Ed.), *Green Business: Hope or Hoax*? Philadelphia, P.A, New Society Publishers, 1991.

Pope, Norman, "*Accounting for Pollution Costs*", Master's Dissertation, Texas University, Lubbock, Texas, 1971.

Porwal, L.S., "*Accounting Theory—An Introduction*", Tata McGraw Hill, New Delhi, 1993, p. 320.

Porwal, L.S. and Sharma, N., "*Social Responsibility Disclosure by Indian Companies*". The Chartered Accountants, February, pp. 630-635.

Preston, L.E., "*Research on Corporate Social Reporting: Directions for Development*", Accounting Organizations and Society, Vol. 6, No. 3. 1981, pp. 255-262.

Ramanathan, K.V., "*Towards a Theory of Corporate Social Accounting*". The Accounting Review, July, 1976, pp. 516-528.

Rifkin, Jeremy., "*Entropy: A New World View*", New York, Viking, 1980.

Roberts, Clare, "Environmental Disclosures in Corporate Annual Reports in Western

Europe", *Green Reporting: Accountancy and the Challenge of the Nineties*, edited by Dave Owen, Chapman and Hall London, 1992.

Ryle, M., *Ecology and Socialism*, Radius, London, 1988.

Security Exchange Commission (SEC), *Securities Act Release No. 5170* (July 19, 1971), Quoted in AAA, Annual Report of the Committee of Environmental Effects of Organisational Behaviour.

SEC: *Securities Act Release No. 5366* (April 20, 1973).

Seidler and Seidler, (edited), *Social Accounting Theory: Issues and Cases*, Los Angeles, Melville Publishing Company, 1975.

Sengupta, P.R., "*Pollution Disclosure in India*", The Chartered Accountant, July 1988 pp. 20-22.

Shrivastava, P., "The Greening of Business" in *Business and the Environment: Implication of the New Environmentalism*. edited by D. Smith, Paul Chapman publishing, London, 1993.

Sinha, G.C. (Dr.), *"Social Balance Sheet"* The Management Accountant, April 1988.

Someya, Kyojiro, "Socio Economic Environment and the Function of Accounting—Does Accounting have a Larger role than External Financial Reporting" in Banerjee Bhabotosh (Ed.), *Contemporary Issues in Accounting Research*, Indian Accounting Association Research Foundation, Calcutta 1991.

Spllerberg, I.F., *Evaluation and Assessment for Conservation*, Chapman and Hall, London, 1922.

Spicer, Barry H., "*Investors, Corporate Social Performance and Information Disclosure: An Empirical Study*", The Accounting Review, January 1978, pp. 94-111.

Sullivan, Thomas F.P., *The Greening of American Business*, Government Institutes, Inc. Rockville, Maryland, USA, Sept. 1992; p. 1.

Tinker, A and Lowe E., "*A Rationale for Corporate Social Reporting : Theory and Evidence from Organisational Research*". Journal of Business Finance and Accounting, Spring 1980, Vol. 7, No. 1, pp. 1-17.

Trotman, K.T., "*Social Responsibility Disclosures by Australian Companies*". The Chartered Accountants in Australia, March 1979. pp. 24-28.

Turner, R.K., (ed.) Sustainable Environmental Management: Principles and Practice, Behharen Press, London, 1988.

Weisner, Pat., "*The Business issues of the '90s*" Colorado Business Magazine, Vol. 17, No. 7, July 1990, p. 8.

Wicks, Clive, "Business and the Environmental Movement" in *Green Reporting: Accountancy and the Challenge of the Nineties*, edited by Dave Owen, Chapman and Hall, London, 1992.

Williams, Caldwell and Needless, "*Reporting Pollution Control Costs in Annual Reports*", Unpublished Paper, May 1972.

22

Environmental Evaluation of Economic Growth: An Agenda for Change

P. Khanna and P. Ram Babu

Sustainable development is a process in which the exploitation of resources, the direction of investments, the orientation of technological development, and institutional changes are all made consistent with future as well as present needs. The premises for sustainable development are:

- Symbiotic relationship between consumer human race and producer natural systems,
- Compatibility between ecology and economy.

The preconditions to sustainable development are:

- Equity and social justice,
- Endogenous choices,
- Economic efficiency,
- Ecologic harmony.

Reprinted with permission from Yojana, 41:8 August 1997.

The present day environmental problems are not so much due to the lack of Governmental thrust as to the direction of its efforts resulting in legalistic, sectoral, media-specific, repair-oriented environmental planning and management that overlook the interactive nature of our common environmental and development concerns. The agenda for change thereby relates to the restructuring of economy based on ecological principles.

The available indicators of growth do not provide environmentally relevant information about the structure of economy. This paper analyses the economic indicators in vogue, and presents a schematic for the incorporation of environmental and resource degradation costs in economic accounting process. The analysis of Indian economy during 1980-95 is also presented to illustrate the utility of the approach in delineating policies for discernible movement towards sustainable development in India.

ECONOMIC INDICATORS

The national income and output account have traditionally served two main purposes:

- measure of level, extent and nature of economic activity;
- delineation of the factors of production, and indicators of living standards.

The links between the measures of income and human well-being (or welfare) are questionable as the income accounts do not, for example, reflect the equity in quality of life, ecological loading and environmental degradation; all of which are relevant to human welfare. Also conspicuous by their absence in accounting procedures are the unrecorded production of goods in informal (unorganized) sectors; resource depletion in building the economy; deterioration in working, living and environmental conditions; and losses due to accidents.

As the gap between improved economic growth and deteriorating quality of life began to widen in 1970s, criticism against GDP as the most important economic indicator has been more articulate, although the critique against GDP as the sole measure of welfare is as old as the origin of GDP itself, ranging from Boulding in 1950s to Daly in 1970s. The latest Club of Rome report, while criticizing the measure of GDP, presents new measures of economic welfare and quality of environment and calls for redirection of economy.

A measure of economic welfare should reflect besides 'classical' material welfare, as described by System for National Accounts (SNA), the following aspects:

- non-market production;
- various parts of production that are not addressed to consumption, but are needed to repair damages caused by the economic system itself (defensive costs);
- environmental damage that is not 'repaired';
- reduction in future welfare caused by production/ consumption today;
- question of (income) distribution.

The value addition from economic activities arising out of the ameliorative investments in environmental quality caused by the economic development itself needs to be subtracted from the GDP to obtain an indicator of positive growth in economy. The items for subtraction from GDP include:

- Costs of unsustainable cultivation of soil
- Loss of natural areas
- Costs of air pollution control
- Costs of noise pollution control
- Costs of water pollution control
- Long-term environmental damages
- Defensive environmental expenditures
- Defensive health expenditures
- Defensive societal costs
- Future reduction in economic welfare
- Depletion of non-renewable resources.

The Chapter 8, Agenda 21 of the Earth Summit calls the Governments to:

> ..... expand existing system of national economic accounts in order to integrate environmental and social dimensions in accounting framework, including atleast satellite systems of natural resources in all member States (8.42)..........'

The statistical offices are responding slowly to this commitment. With the latest revision of the system of national accounts (SNA), the UN is aiming at a paradigm of integrated environmental and economic accounting (SEEA) with a framework to compile environmental data, be it physical or monetary, in a form consistent with the SNA structure. This 'satellite approach' towards environmental accounting was developed jointly by the UN Statistical Division and the World Bank, and pioneered by two country studies in Mexico and Papua New Guinea. Within the SEEA, two measures, viz. depletion of natural resources and environmental costs are included that allow the calculation of 'environmentally adjusted net domestic product' (EDP). The imputation follows the restoration/avoidance cost approach. However, the data could also be organized to compute a set of indicators including measures of Need for Structural Adjustment, Cleaner Technology Options, Substitution of Non-renewable Resource base with Renewable Resources, Weak Sustainability, and Gross Ecological Product (Exhibit I) that could provide a compass for internalizing environmental concerns in the process of socio-economic decision making.

After a decade of research on inclusion of such concerns into national accounts; two broader approaches have emerged:

- Resource and environmental accounts in non-monetary units either to accompany conventional accounts, or to appear separately as satellite accounts.
- Resource use and environmental damage monetization to adjust conventional GNP/GDP measures.

While there is no international agreement on the purpose or the utility of green accounts, different countries have taken recourse to environmental accounting with a view to:

—seeking an indicator of sustainability;

—pursuasive purposes;

—design of environmentally benign sectoral policy responses.

One dimensional sustainability indicators such as the Gross Ecological Product (GEP), or the Environmentally Adjusted Net

EXHIBIT I

Measures of Sustainable Development

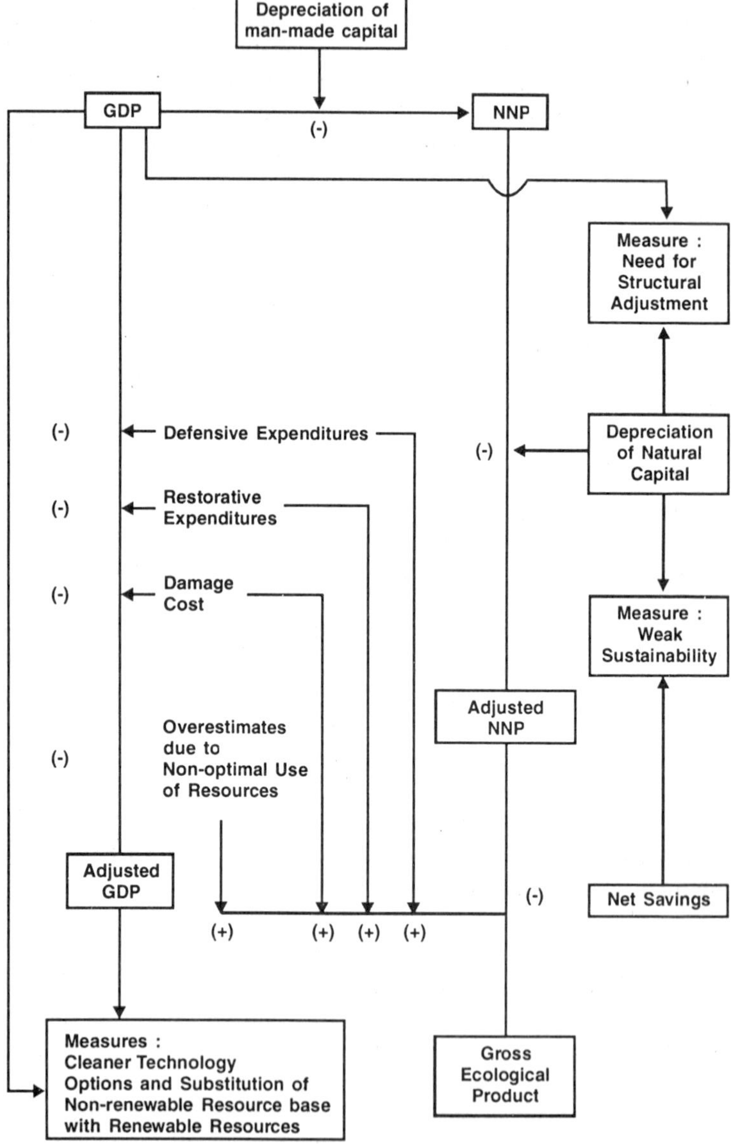

Domestic Product (EDP) provide the extent of environmental and resource degradation unaccounted for in the economic accounts, and their time series reveals the direction in which the economy is movinvg viz. towards sustainability or away from sustainability. The computation of Gross Ecological Product or Environmentally Adjusted Net Domestic Product involves environmental valuation which is ridden with uncertainties and controversies between ecologists and economists. The second indicator, viz. growth of economy towards/away from sustainability, is amenable to computation with recourse to estimation of relative sectoral growth rates, resource use, and emission intensities thus minimizing uncertainties.

NATURAL RESOURCE ACCOUNTS

The Integrated Economic and Environment Accounting framework for India has been a subject of debate since the last decade. A pilot multi-institutional study, sponsored by the Ministry of Environment and Forests in 1995 and coordinated by the National Environmental Engineering Research Institute (NEERI), aims at evolving Natural Resource Accounting framework for India, and its illustration in Yamuna river sub-basin. Another study, covering the whole country, aiming at the estimation of the extent of environmental and resource degradation during the last decade and a half, as an instrument for alerting the decision makers, and emphasizing the significance of environmentally benign paradigms of socio-economic development leading to equity in quality of life, and minimal ecological loading and environmental degradation has also been carried out by NEERI. The salient finding of this study are presented here.

The economic and environment accounts for India, delineated in Exhibit II, include:

- degradation of air quality, with concomitant health and ecological damages;
- unsustainable utilization of ground water resources, and pollution of surface with bodies;
- degradation of land mass due to erosion, salinity, and water logging
- degradation of forest cover.

The issues related to the loss of biodiversity, flora and its crown

EXHIBIT II

Integrated Environment and Economic Accounts (1980-90, 1991-95)

Item	*Change during (1980-90) (Rs. crore)*	*Annual Growth Rate*	*Change during (1991-95) (Rs. crore)*	*Annual Growth Rate*
Economic Accounts				
GDP	+2,02,354	+5.66% (Without accounting for environmental degradation)	+87,721	+4.43% (Without accounting for environmental degradation)
Environment Accounts				
• *Air Environment*				
Damage to health and ecology due to Air Pollution	-26,772	-11,308		
• *Water Environment*				
Ground water mining				
—Quantitative decline	-96,900	48,877		
—Quality degradation	-24,985	-13,386		
Surface water				
—Pollution avoidance cost	-1,014	-512		
• *Land Degradation*				
Productivity losses due to land degradation	=1,38,750	-61,768		
Land rejuvenation cost	-24,000	-10,668		
• *Forest Cover Decline*				
Loss of services/values	-2,704	-1,337		
Total environmental and ecological damage costs	-3,15,125	-1,47,856		
Adjusted growth in GDP (Accounting for Environmental degradation)	-1,12,771	-4.92%	-60,135	-4.74%

Sources of data:

1. National Accounts Statistics, 1995, Central Statistical Organization.
2. India's National Income Statistics, October, 1996, Centre for Monitoring Indian Economy.

density, fresh water aquatic ecological resources, coastal and oceanic resources, and non-renewable energy resources are not included in the calculations. The accounting period (1980-95) is chosen in view of the availability of data on resource degradation.

Economic Accounts

The country's Gross Domestic Product (GDP) at constant prices increased during 1980-90 by Rs. 2,02,354 crore, thus registering a growth of 73.5%; while the increase during 1991-95 was Rs. 87,721 crore, a growth of 21%.

Environmental Accounts

Air Environment

The combined damages to human health and ecological functions of vegetation due to air pollution was Rs. 26,772 and Rs. 11,308 crore during 1980-90 and 1991-95, respectively.

Water Environment

The ground water mining has caused quantitative and qualitative impairment to the ground water resource. The avoidance costs of ground water decline (mining) and degradation are estimated at Rs. 96,900 and Rs. 24,985 crore, during 1980-90; and Rs. 48,877 and Rs. 13,386 crore during 1991-95, respectively. The avoidance costs of surface water pollution are estimated at Rs. 1,014 crore and Rs. 512 crore during the same periods.

Land Environment

The productivity and ecological function losses due to land degradation, and cost involved in rejuvenation and reclamation of degraded landmass during, 1980-90, and 1991-95 are estimated Rs. 1,38,750 and Rs. 24,000 crore; and Rs. 61,768 and Rs. 10,668 crore, respectively.

Biological Environment

The forest cover decline during the periods 1980-90 and 1991-95 has resulted in loss in productivity and ecological services offered by the forest amounting to Rs. 2,704 crore; and Rs. 1,337 crore, respectively.

The salient issues in computation of adjusted growth in GDP, as delineated in Exhibit II, are:

- Health damage costs due to air pollution are based on population exposure, and morbidity and mortality figures that are attributable to respiratory ailments.
- Ecological damage costs due to air pollution include vegetation loss, loss to ground water recharge, and soil erosion.
- Losses due to quantitative decline and quality degradation due to excessive withdrawal of water, and groundwater contamination are estimated.
- Avoidance costs for surface water contamination from domestic and industrial wastewaters are estimated.
- Costs associated with land degradation are productivity losses of land mass under cultivation by assuming aggregate cropping patterns. A period of three years is assumed for land rejuvenation through soil erosion control programme in watersheds.
- Loss of forest services/values is estimated based on the changes in forest cover with recourse to the guidelines of the Ministry of Environment and Forests on benefit—cost evaluation of projects involving diversion of forest land mass.
- Biodiversity losses are not included in the estimation of total environmental and ecological losses.
- Monetized value of natural resources used for growth in GDP has not been included in the calculations of Environment Accounts.

The annual growth rates of gross domestic product and environmentally adjusted net domestic product for some other countries is delineated in Exhibit III.

POLICY IMPLICATIONS

The following observations ensue from Exhibit II with relevance to policy formulations:

- The environmental damages have increased during the

EXHIBIT III

Annual Growth Rates of Gross Domestic Product (GDP) and Environmentally Adjusted Domestic Product (EDP) during 1980-90 and 1991-95 for different Countries

Country	*Rate of growth during 1980-90 (%)*		*Rate of growth during 1991-95 (%)*		*Remarks*
	GDP	*EDP*	*GDP*	*EDP*	
India	5.66	-4.92	4.43	-4.74	The resources and environmental degradation due to informal sector and consumption activities are also included.
Papua New Guinea	4.52	4.32	NA	NA	Only the use of resources and environmental services utilized in production activity is accounted.
Austria	2.26	-0.70	1.0	0.41	The 1991-95 figures are obtained using the data upto 1993. The growth rate figures corresponding to EDP are that of Index of Sustainable Economic Welfare (ISEW).

NA—Not Available

Note: For other countries relevant time series data are not available. The integrated environmental and economic accounting for Mexico in the year 1986 reveals that the contributions of forestry and oil sectors were 0.54 and 3.50 to GDP; and -0.08 and -0.20 to EDP. The capital output ratio for the economy dropped from 37.05 to 9.69 after incorporation of corrections for resources and environmental degradation which constituted 44.8% of GDP.

period spanning 1991-95 in comparison to the period between 1980-90.

- The economic activity during 1991-95 is characterized by larger growth in environmental costs related to water and air pollution, whereas the land degradation and forest cover decline continued with the trends observed during 1980-90.

- The environmental damage to landmass continued to outpace improvements in agricultural production. This combined with the loss of cultivated land for urban expansion seriously jeopardises country's food supportive capacity.

The analysis brings out the crying need for pursuing an agenda for ecologic modernization in the process of economic liberalization. Solely purusing the targets of growth in GDP could lead the economy away from sustainability compromising the growth of future generations.

ECOLOGIC MODERNIZATION

Ecologic Modernization aims at raising the levels of both ecologic and economic efficiency by increasing material and energy effectivity in production and consumption processes in order to minimize the expense on environmental protection while keeping the cost of natural resource exploitation within acceptable limits. In effect, ecologic modernization aims at restructuring of economy based on ecologic principles. A few examples of ecologic modernization are outlined below as an illustration:

- *Manufacturing sector:* transition to production processes which save or recycle raw materials and energy, substitution of ecologically harmful with harmonious products, application of biotechnology for substitution of non-renewable resource base, with renewable, carrying capacity based planning of industrial estates, ecological grouping of industries.
- *Energy sector:* rational use of primary energy, greater use of regenerative energy sources, decentralization of supply, improvement in combustion processes.
- *Agriculture sector:* eco-cultivation and biotechnological improvements, promotion of organic manures and biocides, development of land-use plans compatible with species and ecosystem types.
- *Construction industry:* use of renewable and environmentally compatible building materials, land and energy saving designs, labour-intensive designs.

- *Transport sector:* reduction in specific energy consumption of motor vehicles, reduction in total number of motored kilometers, provision of efficient public transport systems.

The socio-cultural roots of our present environmental crisis lie in the paradigms of scientific materialism and economic determinism which fail to recognize the physical limits imposed by ecological systems on economic activity. The economies must expand within ecosystem which have limited regenerative capacities. Contrary to the neoclassical theory of continuous material growth, economic activities directly undermine the potential for development through over-exploitation of natural resources, and indirectly compromise future production through the discharge of residuals. The entrenchment with quantitative growth as a major instrument of social policy is thus quite paradoxical.

It is this concern that warrants a country wide debate on the present and future scenarios of India's economy with a view to delineating strategies for urgent policy shifts.

23

System of National Accounts and Environmental Accounting

A.C. KULSHRESHTHA AND GULAB SINGH

Accounting of Natural Resource, i.e. incorporating environmental critieria in economic analysis is gaining importance day by day in the entire world and the environmentalists are particularly concerned about the depletion and degradation of natural resources as also about the harmful impacts of economic activities. A number of groups have been constituted both at national and international levels to look into the problems arising out of unmindful handling of natural resources and a few countries are already preparing the Integrated Accounts taking into account the depletion, degradation and other harmful impacts of economic activities. Presently, the economic development of a country or a region is generally expressed in terms of the growth indicated by national income accounts which present systematic statistical statements reflecting the value of final goods and services produced in the economy during a particular year including change in stock. The System of National Accounts (SNA) takes into consideration only the economic activities within the specified

This paper first appeared in the Management Accountant, October 1998, pp. 748-756. Reprinted with permission.

production boundary and records only man-made assets as productive capital. The capital which is used up is written off as consumption of fixed capital against the value of production.

Present System of National Accounts does not keep proper account of natural assets namely land, minerals, water and forest in the sense that they are included in the SNA boundary insofar as they are under the effective control of any institutional unit. The cost of their use is not explicitly accounted for in production cost. It is well known that these natural resources are much more important than the man-made resources since once used up these will not be available to future generations for further production. Day by day the need to have a set of accounts which provide information about the utilisation of these natural resources in a proper way is being increasingly felt and a number of countries have hence initiated to start with on an experimental basis, the exercises of preparing estimates which take into account the details of utilistaion of the Natural Resources. The conventional system of national accounting suffers from two drawbacks—one, the neglect of new scarcities of natural resources which threaten the sustained productivity of the economy and two, the degradation of environmental quality, mainly from pollution and consequential effects on human health and welfare. In addition, some expenditures for maintaining environmental quality are accounted as increases in national income and product, despite the fact that such outlays could be considered as a maintenance cost to society, rather than social progress. In Natural Resource Accounting attempt is made to convert these impacts in monetary terms. Thus the Natural Resource Accounting when properly done may present a completely different picture of the economy in the sense that an economy which might have been considered as a progressive economy on the basis of National Accounts may turn out to be an economy with zero or negative growth if the depletion, degradation and other harmful impacts of economic activities are taken into account.

National Accounts Statistics (NAS) in India as on date are compiled following the guidelines set out in the United Nations System of National Accounts (SNA), 1968. Details of data sources used and methodology adopted are given in the Central Statistical Organisation (CSO) publication, *National Accounts Statistics: Source and Methods,* 1989. India is in the process of revising its national accounts series from the present base 1980-81 to 1990-91 and also simulateneously to adopt to, to the extent feasible, the revised System of National Accounts, 1993 (1993 SNA), prepared under the auspices of the Inter Secretariat Working Group on National Accounts comprising Commission of the European Communities—Eurostat, International Monetary Fund, Organisation for Economic Co-operation and Development, United Nations and World Bank. The revised System is a comprehensive, consistent and flexible set of macro-economic accounts intended to meet the needs of government and private sector analysts, policy-makers and decision-takers. The 1993 SNA embodies the results of harmonising the SNA with other international statistical standards. It recommends preparation of sequence of accounts comprising Current and Accumulation accounts.

PRODUCTION BOUNDARY OF 1993 SNA

The production boundary of the System of National Accounts defines the range of economic activities recorded in the production accounts. Only the activities falling within the production boundary are considered to create output and value added. The production boundary not only defines what is production and what is not, but also determines indirectly what is income, consumption, investment etc. For example, only the output generated within the production boundary could be consumed, invested, exported, etc. In general terms production may be described as an activity in which an enterprise uses inputs to produce outputs. Thus purely natural process, for example, the unmanaged growth of fish stocks in the international waters is not production, whereas the activity of fish farming is production. The production boundary of the 1993 SNA includes the following:

(i) The production of all individual or collective goods or services that are supplied to units or intended to be so supplied, including the production of goods and services used up in the process of producing such goods and services;

(ii) Own-account production of all goods that are retained by their producers for their own final consumption or gross capital formation;

(iii) Own-account production of housing services by owner-occupiers (ownership of dwellings) and of domestic and personal services produced by employing paid domestic staff.

The 1993 SNA includes the production of all goods within the production boundary. At the time the production takes place it may not even be known whether, or not, in what proportions, the goods produced are destined for the market or for own use. With regards to own account production of goods by households, the 1993 SNA has removed the 1968 SNA limitations which excluded the production of goods not made from primary products, the processing of primary products by those who do not produce them and the production of other goods by households who do not sell any part of them on the market.

Towards as effort to compile the Environmental Economic Accounts, cost of use of all assets are to be accounted for which conventionally are not within the SNA asset boundary. In this context it is important to first describe the asset boundary of 1993 SNA.

ASSETS BOUNDARY OF 1993 SNA

Assets an defined in the 1993 SNA are entitites that must be owned by some unit or unit(s) and from which economic benefits are derived by their owner(s) by holding or using them over a period of time. Every economic asset must function as a store of value that depends upon the amounts of the economic benefits that its owner can derive by holding or using it. With regard to the classification of assets, the 1993 SNA distinguishes at the first level of the classification between non-financial assets and financial assets/ liabilities. Within non-financial assets, it distinguishes between produced and non-produced assets and within each of these between tangible and intangible assets. Produced assets are defined as non-financial assets that have come into existence as outputs from processes that fall within the production boundary of the system. This includes not only tangible fixed assets but also intangible fixed assets such as mineral exploration, computer software, entertainment, literary or artistic originals etc. Non-produced assets are defined as non-financial assets that have come into existence in ways other than the process of production. This includes tangible non-produced assets like land, sub-soil assets etc. and intangible non-produced assets like patented entitites, leases and other transferable contracts, purchased goodwill and other intangible non-produced assets.

Gross Fixed Capital Formation

Gross Fixed Capital Formation (GFCF) is measured by the total value of a producer's acquisition, less disposal, of fixed assets during the accounting period plus certain additions to the value of non-produced assets realised by the productive activity of institutional units. Fixed assets are those tangible or intangible assets produced as outputs from processes of production that are themselves used repeatedly or continuously in other processes of production for more than one year.

There is substantial diversity in the different types of gross fixed capital formation that may take place. The following main types may be distinguished:

(a) Acquisitions, less disposals, of new or existing tangible fixed assets, subdivided by types of asset into:
   (i) Dwellings;
   (ii) Other buildings and structures;
   (iii) Machinery and equipment;
   (iv) Cultivated assets—trees and livestock—that are used repeatedly or continuously to produce products such as fruit, rubber, milk etc.,

(b) Acquisition, less disposal, of new and existing intangible fixed assets, sub-divided by type of assets into:
   (i) Mineral exploration,
   (ii) Computer software,
   (iii) Entertainment, literary or artistic originals,
   (iv) Other intangible fixed assets,

(c) Major improvements to tangible non-produced assets, including land,

(d) Costs associated with transfers of ownership of non-produced assets.

Various components of acquisitions and disposals of mixed assets, as referred to in categories (a) and (b) above, are listed below:

(i) Value of fixed assets purchased,
(ii) Value of fixed assets acquired through barter,
(iii) Value of fixed assets received as capital transfers in kind,
(iv) Value of fixed assets retained by their producers for their own use, including the value of any fixed assets being produced on own account that are not yet completed or fully mature, less
(v) Value of existing fixed assets sold,
(vi) Value of existing fixed assets surrendered in barter,
(vii) Value of existing fixed assets surrendered as capital transfers in kind.

Acquisition of new assets covers not only complete assets but also any renovations, reconstruction or enlargements that significantly increase the productive capacity or extent the service life of an existing asset. In recognition of the newly increased capacity or newly extended service life, these improvements are treated as part of acquisitions of new assets even though physically they function as part of existing asset. Items (v), (vi) and (vii) above include disposals of assets that may cease to be used as fixed assets by their new owners: for example, vehicles sold by businesses to households for their personal use or assets that are scrapped or demolished by their new owners.

The 1993 SNA includes a third category of capital formation called "aquisitions less disposals of valuables". Valuables are defined as goods of considerable value that are not used primarily for purposes of production or consumption but are held as stores of value over time. The economic benefits that valuables bring are that their values are not expected to decline relatively to the general price level. They consist of precious metals and stones, jewellery, works of arts etc.

Not all GFCF consists of acquisition less disposals of fixed assets. It is, therefore, convenient to describe the other components of gross fixed capital formation i.e., major improvements and costs of ownership transfers, as these may involve any type of assets. Tangible non-produced assets are natural assets and these are available to the humankind as natural resources. Classification of natural resources has been dealt with in literature (e.g., Judith Rees (1985).

In 1993 SNA the environmental analysis has been dealt with as a functionally oriented satelitte account. The main central accounting framework has not been disturbed. A System of Environmental

Economic Accounts (SEEA) has been developed by the UN and presented in the handbook *Integrated Environmental and Economic Accounting*. There are three approaches to Environmental Accounting, namely, (i) Natural Resource Accounting in Physical Terms, (ii) Environmental Accounts in Monetary Terms, (iii) Welfare and Similar Approach as elaborated below:

Natural Resource Accounting in Physical Terms

This approach focuses on the physical asset balances, i.e. Opening and Closing Stocks and changes therein of materials, energy and natural resources. It may also include changes in environmental quality of natural assets in terms of environmental (quality) indices. The SEEA which shows the links between physical and monetary accounts includes natural resource accounts as a module.

Environmental Accounts in Monetary Terms

This approach identifies the actual expenditure on environmental protection and deals with the treatment of environmental cost of natural and other assets caused by production activities in the calculation of net product. This approach is generally more limited in coverage than physical resource accounting. In monetary approach the GDP is adjusted for selected environmental costs like the cost of oil depletion, deforestation, depletion of fish stock and cost of soil erosion etc. The measurement of such costs is difficult, but considerable research work has been done in this direction, for example, Parikh (1991), Forsund and Strom (1989), Johansson (1989), Repetto (1989) and Tietenberg (1988).

Welfare and Similar Approach

This approach deals with the environmental effects borne by individuals and by producers other than the ones causing these effects. The effects on individuals and on others (that the producers) may be much larger than the cost caused and do not affect net product but rather net income through transfers of environmental services. The approach considers free environmental services provided by nature to producers and consumers and the subsequent damages borne by them. The environmental services provided free and the damages borne are considered as transfers by and to nature which increase or decrease environmentally adjusted net national income.

The General Framework of Environmental Accounts and the SNA

Environmental analysis can be dealt with as a functionally oriented satellite account or in the context of a broadened framework by amending various concepts of the SNA to respond to the growing concern of incorporating environmental criteria in economic analysis. In the integrated economic and environmental satellite accounts, SNA aggregates are amended to treat natural resources as capital in the production of goods and services, to record the cost of using i.e. depleting and degrading those resources and to record the implicit transfers needed to account for the imputed cost and capital items. Table 1 presents the structure of SNA as amended to include the environmental accounts. The flow and stocks of the SNA are shown in the shaded portion and the non-shaded portion shows the additional elements that are needed to supplement the SNA concepts for the purpose of integrated environmental account. The use of the SEEA framework to present environmental accounting and the relationship to the SNA is convenient as the SEEA has been developed in immediate relationship to the SNA so that its concepts and classifications are linked to those of the SNA. The environmental cost and capital elements inclued in the SEEA can be interpreted in the physical terms as well as in monetary terms. However, in view of the controversial issues surrounding valuation of the environmental cost and capital much caution need to be exercised in the use of these elements in monetary terms and in the corresponding derivation of environmentally adjusted aggregates.

1993 SNA Framework

The columns of the Table 1 related to flows are a column (1) for production, covering output (P), intermediate consumption (C), consumption of fixed capital (CFC) and the net domestic product (NDP); a column (2) for the rest of the world, which includes exports (X) minus imports (M) and a column (3) for final consumption. The rows of the table referring to SNA flows are a row (ii) for supply, including outputs and imports; a row (iii) for economic uses, including elements of intermediate cosumption, exports, final consumption and gross capital formation (I_g); a row (iv) for CFC and, finally, a row for (v) NDP which presents the elements that define the national accounts identity between NDP and the expenditure categories. The SNA column (4) for asset balances of produced assets includes the opening and closing stocks of produced assets ($KO_{p.ec}$ & $K1_{p.ec}$) and the elements explaining the changes between the two i.e., net capital formation ($I=I_g$—CFC) holding gains/losses on produced assets ($Rev_{p.ec}$) and other changes in volume of produced assets ($Vol_{p.ec}$). The asset balances in the SNA area cover all economic assets, and therefore, include the assets covered by column (5) for non-produced natural assets. The elements in this column, however, do not figure in the calculation of NDP as all changes in non-produced natural assets between opening and closing stocks ($KO_{np.ec}$ & $K1_{np.ec}$) are explained in the SNA as holding gains/losses ($Rev_{np.ec}$) and other changes in volume of assets ($Vol_{np.ec}$).

SNA Framework extended to Environmental Accounts

The non-shaded areas of Table 1 include the additional elements that are needed to supplement the SNA concepts with data in physical terms on environmental cost and capital. There are two types of additional elements. The first group is included in an additional column (6) which records the effects of economic activities on non-produced natural assets such as air, water and virgin forests that are not included as economic assets in the SNA. The second group of elements is included in two additional rows (vi-vii) that include elements for the use of non-produced natural assets by depletion and degradation, and for other accumulation of non-produced natural assets, which cover the transfer of natural assets to and between economic uses. Another row (viii) is included to derive an environmentally adjusted net domestic product (EDP) and other environmentally adjusted concepts. This row is only relevant in the case of monetary environmental accounting when additional SEEA elements are specified in value terms.

In row (vi) related to the use of non-produced natural assets, an additional element Use_{np} has been included in the column for production. This reflects the use of non-produced natural assets in production; it is sum of counterpart items in column (5) and (6) representing respectively; the use of non-produced natural assets that are economic assets in the SNA sense, i.e. ($-Use_{np.ec}$) and the degradation of other natural assets that are not economic assets i.e. ($Use_{enp.env}$). The use of non-produced economic assets ($-Use_{np.ec}$) includes the depletion of minerals, extraction of timber from forests that are economic assets and the effects on productivity of those forests and agricultural land of soil erosion, acid rain etc. *The deteriorating effects of air pollution on buildings and structures and the effects of soil erosion on roads and other degrading effects on produced assets are not included as*

TABLE 1

The Basic Structure of SEEA

		Economic activities		*Economic Assets*		*Environment*	
		Production	*Rest of World*	*Final Consumption*	*Produced assets*	*Non-produced assets natural assets*	*Other non-produced*
	1	2	*3*	*4*	*5*	*6*	*7*
(i)	Opening stock of assets				KOp.ec (Opening stock of produced assets)	KOnp.ec (Opening stock of non-produced natural assets)	
(ii)	Supply	P (Output)	M (Imports)				
(iii)	Economic uses	C_I (Intermediate Consumption)	X (Exports)	C (Final Consumption	I_g (Gros capital formation)		
(iv)	CFC	CFC			-CFC		
(v)	NDP	NDP $=P.C_I$-CFC	X-M (Net exports)	C (Final Consumption)	I $=I_a$-CFC (Net capital formation)		
(vi)	Use of non-produced natural assets	Use_{np} (Use on non-produced natural assets in production)				$-Us_{cnp.ec}$ (Use on non-produced economic natural assets	$-Us_{cnp.env}$ (Degradation of non-economic natural assets)

(Contd.)

TABLE 1 *(Contd.)*

	1	2	3	4	5	6	7
(vii)	Other accumulation of non-produced natural-assets					Inp.ec (Change in stock of non-produced economic assets)	-Inp.env (Reuction in natural assets other than assets)
viii)	Environmentally adjusted aggregates in monetary environmental accounting	EDP $=C+(A_{p.ec}$ $+A_{np.ec})$ $+A_{np.ec})$ -Anp.env + (X-M)	X-M (Net exports)	C (Final Consumption)	$A_{p.ec}$ (Net accumulation of produced assets)	$A_{np.ec}$ (Net accumulation of non-produced economic assets)	$-A_{np.env}$ (Net accumulation of other non-produced natural assets)
ix)	Holding gains/losses				$Rev_{p.c}$ (Holding gains/losses of produced assets)	$Rev_{np.ec}$ (Holding gains/losses of non-produced natural assets)	
(x)	Other changes in volume of assets				$Vol_{p.ec}$ (Other changes in volume of produced assets)	$Vol_{np.ec}$ (Other changes in volume of non-produced natural assets)	
(xi)	Closing stock of assets					$Kl_{p.ec}$ (Closing stock of produced assets)	$Kl_{np.ec}$ (Closing stock of non produced natural assets)

they are assumed to be reflected in the CFC. The use of non produced natural assets that are not economic assets ($Use_{np.env}$) covers the non-sustainable extraction of fishstock from oceans and rivers, extraction of firewood and lumber from tropical and other virgin forests of hunting animals living in the wild and also the effects of emission of residues on the quality of air, water, fishstock, wild forests and the effect of other economic activities (recreation, agriculture, transport etc.) on ecosystem and species habitat.

Other accumulation in row (vii) records in physical or monetary terms the transfer of natural assets to economic uses as a change in the stock of non-produced economic assets ($I_{np.ec}$). The counter part of this increase in economic asset is the reduction of natural assets other than economic assets ($-I_{np.env}$). Thus $I_{np.ec}$ would include the transfer of land to economic uses, the net additions to proven mineral reserves, the conservation of wild forests to timber tracts or agricultural land and the conversion of fishstocks to economic control. If the deterioration takes place at the same time that natural assets are incorporated as economic assets, the deterioration is not recorded in the other accumulation row, but is included as part of use of natural resources. If this deterioration takes place before the transfer, it is recorded as uses of natural resources that are economic assets ($-Use_{np.ec}$). As the elements of row (vi) for use (i.e. depletion or degradation) of non-produced natural assets, and row (vii) for other accumulation are included in the SNA in other volume changes, the content of other volume changes is reduced in the SEEA as compared with the SNA. *If the additional SEEA elements are valued in monetary terms, the incorporation of the use of non-produced natural assets (Use_{np}) as additional cost in the column for production results in an EDP, presented in row (viii), which is lower than NDP. The elements in row (vii) for other accumulation do not effect EDP.*

Corresponding to the monetary valuation of the additional SEEA elements, on the expenditure side a new concept called net accumulation is introduced in the SEEA to replace net capital formation in the SNA. It is presented in row (viii), separately for produced assets ($An_{p.ec}$), and other natural assets ($A_{p.ec}$), non-produced economic assets ($A_{np.env}$). For produced assets, it is same as net capital formation (i.e. $A_{p.ec}=I$), for non-produced economic assets, it reflects the net effects of negative depletion and degradation and positive additions of natural assets that are transferred to economic uses (i.e. $A_{np.ec}=- Use_{np.ec} +I_{np.ec}$). For natural assets other than economic assets it could be considered as the economic valuation of the impact of economic activities on the environment. It is thus sum of negative depletion and degradation effects ($-Use_{np.env}$) and negative effects of incorporating natural assets as economic assets (i.e. $-A_{np.env} = -Use_{np.env} - I_{np.env}$).

If net accumulation replaces net capital formation when the additional SEEA elements are valued in monetary terms, the national accounts identity between NDP and final expenditures changes. In the SNA this identity, as reflected in row (v) of Table 1 is:

$$NDP = C + I + (X-M) \ldots\ldots \quad (1)$$

and EDP is derived from column (1) is as follows:

$$EDP = NDP - Use_{np} \ldots\ldots \quad (2)$$

If net capital accumulation in economic assets replaces net capital for mation (I), the identity as reflected in row (viii) becomes:

$$EDP = C + (A_{p.ec} + A_{np.ec}) - A_{np.env} + (X-M) \ldots\ldots \quad (3)$$

Equation (3) can be proved to be equal to equation (1). We know from definition that—

$$Use_{np} = -Use_{np.ec} - Use_{np.env} \ldots \quad (4)$$

Also,

$$A_{p.ec} = I \ldots\ldots \quad (5)$$

and

$$A_{np.ec} = -Use_{np.ec} + I_{np.ec} \ldots\ldots \quad (6)$$

and

$$-A_{np.env} = -Use_{np.env} - I_{np.env} \ldots \quad (7)$$

Replacing LHS of equations (5), (6) and (7) in equation (3) we get

$$EDP = C + (I - Use_{np.ec} + I_{np.ec})$$

$$Use_{np.env} - I_{np.env} + (X-M)$$

$$= C + I + (X-M) - Use_{np.ec} - Use_{np.env} + I_{np.ec} - I_{np.env}$$

Now as $I_{np.ec}$ is the counterpart of $-I_{np.env}$, therefore, last two terms cancel.

Thus,

$$EDP = C + I + (X-M) - Use_{np.ec}\ Use_{np.env}$$

Now using (4), we get

$$EDP = C + I + (X-M) - Use_{np} \text{ or } = NDP - Use_{np}$$

Details of Environmental Amendments to SNA Framework, Concepts and Classifications.

Assets Boundary and Classification

As per 1993 SNA the Natural Assets are included only if they provide economic benefits to the owner/controlling institution. This may mean explicit ownership and/or availability of a market price, such assets are referred to in SNA as economic assets. In SEEA however, the asset boundary has been much widely defined and particularly include all the natural assets some of which may directly help in production while others may be affected by environmental impacts of economic activities. The asset boundary of SNA inclues only economic assets as dealt with in columns (4) and (5) of Table 1. The SEEA asset boundary comprises all natural assets including those that are economic assets and are covered in columns (4) and (5) and other natural assets that are represented by column (6). Table 2 below shows that some assets categories in the SNA and SEEA are identical, some are closely related with different coverage and some are not at all included in the SNA but are included only in the SEEA. The asset categories with an identical coverage are cultivated assets such as orchards and plantations and work in progress on cultivated assets including agriculture crops and livestock, which are treated as inventories in both the systems. Further the subsoil assets category is also identical in both the classifications. Air is one of the assets included in SEEA because it is affected by economic activities but is not included at all in the SNA. Land although covered in both the systems have widely different coverage in the systems in the sense that SNA land includes associated water surfaces such as lakes and rivers and in some instances also ground water. Since the SEEA deals with the effects of economic activities on the quality of water in rivers and lakes as also on the quality and quantity of available ground water it excludes these categories of water resource from land and includes them as sub categories under the category of water. Thus the category of water in SEEA is much broader than the SNA category. The SNA category is mainly restricted to aquifers that are controlled by human activity while the SEEA land includes the ecosystem which are not explicitly included as economic assets in SNA. This classification of natural assets for environmental accounting is required in order to fully cover depletion, degradation and other accumulation, i.e. transfer of natural assets to economic activities. This classification will go whether physical units or monetary values are used for Environmental Accounting.

ENVIRONMENTAL COST

SEEA identifies two types of environmental cost viz. imputed cost for degradation and depletion and actual cost in the form of environmental protection expenses. These costs are explained as under:

The Cost of Use of Non-Produced Natural Assets: this cost has been introduced as an additional cost in SEEA and represents the cost of depletion and degradation in physical terms e.g. quantity of minerals extracted, quantity of timber cut, value of solid, liquid or gaseous wastes generated. Depletion is not restricted to economic assets only as in SNA and may cover depletion of uncontrolled forests for the purpose of firewood or the uncontrolled depletion of fish stocks. In these cases the assets are considered as non-economic in the sense that the exploitation of assets is uncontrolled. In such cases a depletion allowance is calculated in the SEEA except if the depletion in the case of wild biota is within the bounds of natural growth and the depleted assets would be naturally replaced. If expressed in monetary terms the depletion allowances could be considered as payments for the extraction of non produced assets, such as minerals which are in stock either as economic assets or as Natural Assets. Depletion of non produced natural assets is therefore considered in the SEEA as cost and at the same time treated as negative changes in inventories. Degradation effects in physical or monetary terms are included as additional cost elements in physical or monetary terms no matter whether these affect natural assets that are economic assets or other non economic natural assets. The affects of degradation may include the deterioration of natural assets as a result of their use as sink for residues or as a consequence of other degrading activities. The deterioration of produced assets due to degradation is assumed to be reflected in CFC accounted for in the SNA. Environmental degradation by households is treated as a cost of internal household production activities and environmental degradation caused by fixed assets abandoned in the environment at the end of their useful economic life is allocated as cost of degradation to the producers owing those fixed assets. Both adjustments reduce EDP as compared to NDP in monetary environmental accounting. Another convention of SEEA

TABLE 2

Classification of Natural Assets in the SNA and the SEEA

SNA	*SEEA*
Produced Assets	
Fixed assets	
Cultivated assets	
Inventories	
Work-in-progress on cultivated assets	
Non-produced Assets	
Tangible Non-produced Assets	*Tangible Non-produced Assets*
Land including associated surface water	Land including ecosystems
Sub-soil assets	Sub-soil assets
Non-cultivated biological resources	Wild biota
Water resources	Water
	Air

relates to environmental degradation "imported from" or "exported to" neighbouring countries. Net of these is treated as a further environmental cost element.

Environmental Protection Expenses

Some actual expenses are incurred to avoid environmental degradation or to eliminate the effects after degradation takes place. Enterprises explicitly produce such services on a commercial basis. In many instances however, the services are produced as the ancillary activities. The ancillary activities are treated as separate establishment with an identifiable output and intermediate costs. The externalisation of ancillary environmental activities is done in order to measure and evaluate more comprehensively the efforts actually made to combat the environmental degradation and its effects. In general the SEEA considers only those environmental protection expenses which are in immediate response to the effects caused by production. It excludes environmental protection expenditure responding to other environmental impacts that is why household expenses for environmental protection are not considered they respond to impacts borne by households and are dealt with in the SEEA in the same manner as in the SNA. An exception is made for environmental services produced as ancillary activities by government, which are included in the SNA as part of non market output. They are identified in the SEEA as separate establishment.

Net Capital Accumulation: Capital formation in the SNA includes changes in the stock of produced assets used in production that are caused by economic decisions. Capital accumulation in the SEEA includes not only capital formation but also changes in non produced natural assets that are explicitly used in production and are thus also related to economic decision. Such changes include the reductions in the capital stock as a result of depletion and degradation and also the incorporation of natural assets as economic assets and the transfer of natural assets beteween economic uses as a result of economic decisions taken in connection with production activities. Catastrophic losses due to events such as wars and natural disasters are not included in the concept of Net Capital Accumulation unless they are caused in connection with productive activities.

VALUATION

Three alternative valuation methods are available viz. market prices, valuation cost and contingent valuation. In the SEEA the principles of market valuation are the same as in the SNA, they are applied to stocks of non produced natural assets that are economic assets in the SNA, to changes therein as a result of uses of natural assets and also to the depletion of natural assets that are not economic assets such as the extraction of fish from the oceans and firewood from virgin forests. Market valuation however has been more extensively elaborated in the SEEA insofar as natural resources are concerned. The only difference between the SNA and the SEEA is that the use of non-produced natural economic assets is recorded in the SNA as other volume changes while in the SEEA cost of degradation and depletion is reflected in capital accumulation and deducted in the derivation of EDP. Opening and Closing stock of other non produced natural assets which are non-economic assets in SNA are recorded in the SEEA in physical terms. Degradation of air and water is not valued in either system as these are uses of non-produced natural assets that are non economic assets. However while the SNA excludes the SEEA includes in its market valuation all uses of natural assets that are not economic assets like the depletion of fish stock from the ocean, the extraction of firewood from virgin forests as the fish and firewood have a market value. Two valuation methods are used for depletion one is net rent approach which values the units extracted on the basis of the difference between out put and a costs including labour cost and a normal profit margin incurred as a result of depletion. The other is user cost approach which values the units extracted on the basis of only a part of net rent, i.e., that part which if reinvested would generate a permanent income stream equal to the loss of income generation capacity through depletion. SEEA analyses the quality changes in assets, i.e. degradation separately, deducting this value in principle from value added generated and capital accumulation.

The two other valuations used in the SEEA viz. maintenance cost and contingency valuation are only applied to uses and not to stocks of non-produced natural resources. Their use may however lead to a number of valuation inconsistencies within the SEEA.

Endeavour to Compile SEEA in India

The environment statistics is a newly emerging area of importance. The need for development of an appropriate statistical system has been stressed at various forums and has engaged attention of the CSO. The matter has been discussed in the Conference of Central and State Statistical Organisations and the consensus has been that environment being a multidisciplinary concept involving large number of resource elements both natural and man-made, it wuld be necessary to ensure meaningful linkage of statistics from collection, compilation, and analysis through effective participation of concerned agencies at the centre and at the state levels. In 1986, an Inter-Disciplinary Working Group with members from several organisations concerned with environment management was set up under the Chairmanship of Director General, CSO. This group identified the major areas of environmental concern on which data need to be developed and the parameters pertaining to the areas, assigning priorities and suggesting methodology for compilation of data on these parameters. The major areas of concern namely; flora, fauna, atmosphere, water, land and soil and human settlements were considered and current availability of data noted and suggestions made for collection of further data. The group also came out with a provisional list of statistical variables for each of the major areas. Recently a Compendium on environment statistics has been prepared by the CSO. This Compendium has been prepared under the advice of a Technical Committee set up for the purpose. It gives the available statistics/figures at a glance. The Asian Development Bank has provided a token grant for the preparation of the Compendium, holding a national workshop and training of personnel on the various issues. It is proposed to identify the data gaps which exist in the various areas of the environment statistics and statistics for obtaining data through a system.

In the spirit of the recommendations of the United Nations Conference on Environment and Development (UNCED), Ministry of Environment and Forests have sponsored few studies to evolve suitable methodology for compilation of natural resource accounts for natural assets like, water and forests. In fact what is required is the establishment of framework for environmental accounting. Once the framework for environmental accountng has been established, it can also serve in implementing special studies that aim at improving the data contents and analysis of particular sectors of the framework. One type of studies could focus on the asset accounts for indepth studies of particular natural assets. Detailed inventories of natural resources would measure not only asset stocks but also changes therein and their economic and non-economic causes. Examples of such studies are the Philippines Forest Resources Accounts and a more experimental compilation of crude oil and natural gas accounts in Canada. The use of SEEA framework would avoid the risk of non-compatibility with national accounts concepts and procedures which is a major drawback of adhoc studies carried out outside the national statistical services.

References

Central Statistical Organisation (1989): National Accounts Statistics, Sources and Methods. Department of Statistics, New Delhi.

Forsund, Finn R. and Steiner Strom (1988): Environmental Economics and Management : Pollution and Natural Resource: New York.

Johansson, Per-Olvo (1989): Valuing Environmental Damage. Oxford Review of Economic Policy. Vol. 6, No. 1.

Parikh, Kirit S. (1991), Towards a natural resource accounting system. *Journal of Income and Wealth*, 13(2).

Rees, Judith (1985): *Natural Resources: Allocation, Economics and Policy*. Methuen.

Repetto, Robert, *et. al.* (1989): Wasting Assets: Natural Resources in the National Income Accounts. World Resource Institute, Washington, DC.

Geological Survey (1976): Principles of the mineral reserves clasification system of the US Bureau of Mines and the US Geological Survey. *Geological Survey Bulletin*, 140-A.

National Accounts Statistics: Sources and Methods, 1989, Central Statistical Organisation, Department of Statistics, Government of India.

System of National Accounts 1993 (1993 SNA), Commission of the European Communities, International Monetary Fund, Organisation for Economic Co-operation and Development, United Nations, World Bank, Brussels/ Luxembourg, New York, Paris, Washington, DC, 1993.

24

Environmental Audit—A Perspective

NABA KUMAR PATNAIK

Audit is an important Management Information System (MIS) tool, on which many management decisions are dependent. The purpose of any audit is to acquire information and provide additional discipline on the internal processes to validate proper functioning of the specific system. The concept of environmental audit is nothing new as different expressions like "Environmental Surveillance", "Environmental Review", "Environmental Assurance", "Environmental Quality Control" and "Environmental Assessment" have been used by different companies. The USA and UK have introduced their own environmental audit system in 1960 and 1970 respectively.

DEFINITION

Environmental Audit is a very broad term as "environment" in itself comprises of biotic and abiotic componenets as is well defined in the Environment (Protection) Act, 1986: "Environment includes water, air and land and the inter-relationship which exist among and between water, air and land and human beings, other living creatures, plants, microorganism and property". Environmental Audit, therefore, is an independent evaluation of (i) Policy and Principles, (ii) Systems,

(iii) Procedures, (iv) Practices and (v) Performance of elements of an industry relating to environment. It is evident that the scope of environmental audit is much wider than those specified by Indian Regulatory Agencies. It is a management tool comprising a systematic, documented, periodic, and objective evaluation of how well environmental organisation, management and equipment are performing with the aim of helping to safeguard the environment by:

(i) Facilitating management control of environmental practices;
(ii) Assessing compliance with company policies, which would include meeting regulatory requirements".

OBJECTIVES

The objectives of an Environmental Audit are, evaluation of the efficiency and efficacy of resource utilisation (i.e. men, machines and materials); identification of areas of risk, environmental liabilities, weakness in management systems and problems in complying with regulatory requirement, and ensuring the control on waste/pollutant generation.

Environmental Audit has been dealt under the following categories:

Design Specification and Layout

While setting up an industry, adequate provisions are kept in the design, specification and layout to augment the production capacity but corresponding provisions to meet the environmental criteria are often overlooked. Adequate provisions are further necessary to up-grade pollution control measures to meet the future environmental standards that are getting stringent day by day. The audit would help in identifying specific area of concern to meet the future requirement of environmental measures.

Resource Management

The resources include air, water, land, energy and other raw-materials. This has to be compared with the best achievable for the industry in general and through best available Technology. The human resources if exploited to its maximum potential, would be an unenviable asset to the industry. In the present day context, surplus man-power could often be seen hindering the desired progress. Surplus unqualities and inexperienced manpower is diverted to newly created environmental management department. Success of a system depends on the deployment of suitable human resources and their efficient utilisation. The human resource audit would identify the suitability of the staff deployed and their training needs. The audit would provide data to the management on efficient use of the resources per unit production there by reducing resource consumption and minimising the waste. This exercise is more beneficial for mega industries where even slight economy on resources would boost for energy consumption, and is reflected in the Annual Report of the Company.

Pollution Control Systems and Procedures

The objective of conformance audit is to ensure that the systems and procedures governing the environmental activities/operations of pollution control equipments are rightly followed. While the conformance audit would determine whether the systems and procedures are provided, the effectiveness of audit would determine the efficiency of the system in identifying conditions inviting corrective actions in a timely and effective manner. Further, for preparing pollution inventory, all the emission sources and process units need be documented and emission inventory prepared based on the throughout material, usages, plant record and emission.

Emergency Plans and Response/Safety System

The concerned staff is often, baffled for right action during emergency as the emergency plans often remain in the safe custody. The problem becomes acute when new persons are employed/deployed. The review of this system will ensure adequate knowledge and alertness of the concerned staff and readiness of the paraphernalia to face an emergency.

Medical and Health Facilities/Industrial Hygiene and Occupational Health

The productive element of an industry is dependent on its healthy manpower. Healthy mind in a healthy body carves out a healthy outlook of any organisation. The primary facilities to suit the occupational needs of the industry are therefore, more vital. Audit in this regard would provide an insight into the actual requirements to warn suitable orientation of existing facilities.

Information System/Institutional Set-up

The institutional set-up establishes line of authority, responsibility and communication within the organization to ensure smooth functioning. There is a need to evaluate the relationship between various systems that aid one another on reaching common goal and eliminate friction and cross purposes. In view of the Environmental Audit becoming compulsory, a summary of compliance with environmental laws, in the report of Board of Directors would be quite useful.

Inter-relationship of various Environmental Domains

An environmental impact assessment (EIA) is normally carried out prior to setting up an industry. EIA, predicts likely future scenario through superimposition of the impact due to the development activity on the existing environmental conditions. It further suggests the mitigatory measures to ensure the environmental compatibility of the industry. The mitigatory measures complying with conditions are then incorporated in the project design and systems. However the impact on the various domains of environment in reality could be different. In a few cases the impact could be more deleterious than the predictions made on the EIA. This may be due to various extraneous climatological and environmental factors not attributable to the particular industry. The inter-relational environmental audit would evaluate the environmental scenario of the industry in operation; assess the impact of industry; its share on the degradation of environment; validate the findings of EIA Study and suggest improvement in EIA Methodology.

Conformation to Regulatory Requirement

The regulatory mechanism of environmental compliance is gradually becoming more and more comprehensive. The new regulations and standards are being stipulated at such a pace that it would render the existing system archaic. Factory managers more often are unaware of the latest requirement, making the top management/owners vulnerable for prosecution under various environmental acts. The objective of compliance audit is to compare the existing status with the stipulation and standards prescribed by various regulatory agencies and ensure compliance to various regulatory requirements. The regulatory requirements will have to be upgraded each year.

Social Benefits

The local populace often gets alieniated in their own land and develop xenophobia due to setting up of an industry. The social audit would evaluate the impact of an industry in the local populace in their smooth transformation from agrarian environment into industrial environment.

Audit Procedure

For smaller and medium industry, the audit may be performed on a basis so informal that no one including the Manager realizes that there is infact an audit. However, for the larger industries, it has to be well-planned.

The audit should be carried out at a regular interval. The internal audit should be more frequent supplimented by random audit through external agencies which would help identify the training needs of audit staff.

The domains of environmental audit would vary depending on the nature of industry as for example. "Social Benefit Audit" would be more important for an industry requiring large acreage of land; "Emergency Planning and response/safety Audit" would be vital for a chemical industry; and "Resource Audit" would be the focus for an industry with high resource consumption. However, the "Pollution Control System and Procedures" and "Conformation to Regulatory Requirement" would be basic requirements for all industries.

Different types of audit should be conducted by separate/distinct groups in order to maintain absolute objectivity. The report should be based on facts supported by documentary evidence of the site management and should include action plan for corrective measures with specific responsibilities and deadliness. A specific structured questionnaire should be prepared for consistency and follow up action. A follow-up action plan should be drawn up by the site officials in consultation with the auditor for implementing the recommendations of the audit and minimise deficiencies/short commings in the system. The difference of opinion of any, could be sorted out through discussions with the top management.

Audit Personnel

The audit should be carried out by a qualified and experienced personnel who are not directly involved in the day to day affair of the factory. They should however be well-versed with various processes of the industry which would thereby help management to an unbiased evaluation and dissemination of information.

The task of environmental audit is indeed, herculean in a country like India, owing to its wide diversity. It would therefore be prudent that the authorities contemplate setting up of an Institute of Chartered Environmentalist (ICE) on the lines of Institute of Chartered Accountant. A lead could be taken by the national associations like FICCI, CII and ASSO CHEM. The members of ICE could meet challenges of the 21st. century in the liberalised Indian economy and industry. It is the basis for developmental planning whose ultimate aim is social benefits. It would help the industry in assessing the performance of business in terms of utilisation of country's resources.

The Annual Reports of the Indian Corporate Industry generally mention Environmental aspects reflecting the intent of the top management to protect the environment. This could further be made transparent by incorporating Environmental Audit Report in the Annual Report for the information of share holders and general public. The efficient and honest reporting will increase the good-will of the company and the value of shares. Regulatory Agencies should ensure that the excerpts of Environmental Audit be covered in the prospects by the companies making the public aware of the environmental policies of the company and its risk and liabilities.

Audit not only help the industry but, also the nation to develop a sound scientific data base for decision on achievable technologies and standardize the regulatory requirements accordingly. Environmental Audit would gradually pave way for adopting total quality management systems for the industry. The environmental audit similar to the quality control would not add to cost of the product but to value of the product and company.

The term Environmental Audit has been changed to 'Environmental statement'.

The preparation of "environmental statement" has become

mandatory for all Indian industries since 1993. This has necessitated environmental audit for a proper statement to be produced. A proper environmental audit should bring out the following facts:

(i) Proper estimation of inputs of raw materials and output of products. This would give estimate of some of the wastes going into environment as pollutants.
(ii) The efficiency of the control measures that have been adopted, as well as further control measures that may be necessary for control of pollution generated.
(iii) Return from recycle/recovery of waste materials and whether further improvements of materials and energy recovery are possible.
(iv) Collection and maintenance of data for environmental management: whether the present method is satisfactory or improvements needed.
(v) Recommendations for a long term pollution free operation.

Environmental Audit Report for the Financial Year Ending the 31st March

PART A

(i) Name and address of the owner/occupation of the industry, operation or process:
(ii) Date of the last environmental audit report submitted.

PART B

Water and Raw Material Consumption

(i) Water consumption

Process

Cooling

Domestic

Name of Products	Water Consumption per unit of products	
	During the previous financial year	During the current financial year
	(1)	(2)
1.		
2.		
3.		

(ii) Raw Material Consumption

Name of raw material	Name of products	Consumption of raw material per unit of output
	During the previous financial year	During the current financial year
	(1)	(2)

PART C
POLLUTION GENERATED

Parameter as specified in the consent issued		
Pollutants	Quantity of Pollution generated	Percentage of variation from prescribed standard with reasons
(a) Water		
(b) Air		

PART D
HAZARDOUS WASTES

(as specified under Hazardous Wastes/Management and Handling Rules, 1989)

Hazardous Wastes	Total Quantity	
	During the previous financial year	During the current financial year
(a) From process		
(b) From pollution control facilities		

PART E
SOLID WASTES

	Total Quantity	
	During the previous financial year	During the current financial year
(a) From Process		
(b) From pollution control facility		
(c) Quality recycled or recitilised		

PART F

Please specify the characteristic (in terms of concentration and quantum) of hazardous as well as solid wastes and indicate disposal practice adopted for both these categories of wastes.

PART G

Impact of the pollution control measures of conservation of natural resources and consequently on the cost of production.

PART H

Additional investment proposal for environmental protection including abatement of pollution.

REFERENCES

Trivedi, Priya Ranjan; Singh, Uttam Kumar; Sudashan, K. Cherry and Tuteja, Tripat Kaur—International Encyclopaedia of Ecology and Environment; Indian Institute of Ecology and Environment, New Delhi, 1994.

Tondon, B.N. Practical Auditing, S. Chand & Co., New Delhi, 1984.

Das, R.C., Environmental Laws in Relation to Pollution Control—Save Environment Save Yourself, Orissa Environmental Society, Bhubaneswar, 1991, pp. 52-61.

The First Citizen's Report—Centre for Science and Environment, New Delhi, 1987.

The Second Citizen's Report—Centre for Science and Environment, New Delhi, 1985.

Annual Report Tata Iron & Steel Co., Mumbai, 1993-94.

Lekhi, R.K. The Economic Development and Planning, Kalyani Publisher, New Delhi, 1990.

25

Environmental Audit: Green Accounting

NAGESH KINI

India is considered to be one of the few countries in the developing world which have comprehensive regulations pertaining to the environment. However, the administration of pollution control has been dismal due to want of reliable information on the violation and implementation of these regulations. At a seminar on *Environment Management in Industry* organised by FICCI, the Chairman of the Central Board of Pollution Control went on record to declare that the failure of industry to comply with the government's 1993 initiative on Environment Audit will have an adverse impact on the development of an effective and viable legal system for environment protection.

THE BACKGROUND

Only three per cent of industry have submitted environment audit reports. What is worse is that the status of compliance is very low for old plants. In Karnataka, 48 out of 99 units set up before 1991 do not conform to the present environment standards. Compliance

This paper first appeared in the Chartered Accountant, November, 1998, 36-38. Reprinted with permission.

has been low in the leather and sugar industries. The compliance ratio of paper and pulp units and distilleries has been less than 50 per cent. Moreover, pollution by small-scale units goes unreported.

GROWTH OF POLLUTING INDUSTRIES

The growth of petrochemical refineries, thermal power generation, chemicals, food processing, fertilisers, caustic soda, iron and steel has serious implications for the environment. The first three account for over 70 per cent of Foreign Direct Investment. According to a Reserve Bank of India survey on investments, Gujarat accounted for 24 per cent (Rs. 15,031 crore) and Maharashtra for 17.5 per cent (Rs. 10,492 crore) of total industrial investment made in 1994-95. The number of projects in Gujarat was 159, in Tamil Nadu, 252.

A report in *The Times of India* on November 25, 1995, points out that heavy concentration of industrial investment in certain areas of Gujarat has had negative consequences. In Asia's largest industrial estate at Ankleswar in Gujarat, 3,000 chemical units constitute over half of the total units. They reportedly generate 270 ml. of hazardous liquid waste daily, and 50,000 tonnes of hazardous solid waste annually. To add to the problem, multinationals like Shell, ICI, Dupont, and Pculenc, the world's chemical giants, are also investing here.

Objectives

1. Does the organisation have a statement defining its environmental objectives in specific terms? If not, can the objectives be identified? What are they?
2. Are the broad objectives broken own into specific targets, expressed in quantitative terms and formulated for each responsibility centre? Are the tasks and responsibilities clearly listed?
3. Do clear-cut policy guidelines exist in various areas of environmental management?
4. Has the company considered all Regulations, Acts Notifications of Local Authorities, State and Central Governments, Regulatory, Bodies and International Standards in determining the objectives and guidelines and whether a prompt adequate and systematic reporting procedure exists and is actually in operation?

Planning

1. Have short-and long-range plans for environmental management been formulated?
2. Is there an adequate system, actually in operation for formulation, scrutiny, approval, implementation and review of the plans?

Organisation

1. Does the company have a well-designed formal organisational framework covering the persons in charge of environmental management and laying down the lines of authority and vertical and horizontal relationships?

General Opinion

1. What is the opinion of the management, staff, workers, suppliers, business associates, customers, green groups and the general public about the state of affairs of the particular company?

The doubling of capacity utilisation by all industries between 1961-62 and 1991-92 has had significant environmental consequences, especially in Maharashtra, Gujarat, Uttar Pradesh, and Tamil Nadu, where the largest amount of investment was made.

This points to the urgent need for authenticated information on the state of pollution, environment management measures, and implementation of eco-friendly technologies in the country.

AUTHENCITATING ENVIRONMENTAL REPORTS

Whether industry's oft repeated assertions of commitment to environmental protection have actually been translated into action can only be ascertained when their published environment reporting data is authenticated by subjecting it to independent attestation. The Annual Financial Statements of companies have a wider readership than any other publication or document, in as much as it is read by the widest section of users. Being a public document under the statute, it is available to all and easily accessible. This, rather than any other form of compliance report with restricted circulation, is the best mode of presenting environmental disclosures, compliance, and achievements.

The annual accounts are statutorily required to be compiled and published at regular intervals. They have necessarily to be made out in the prescribed format to include disclosure and are submitted to various authorities, shareholders and institutions on pain of penalties and prosecution for non-submission.

All that needs to be done is to include environmental data in the disclosure requirements in Part II of Schedule VI of the Companies Act, which, inter alia, lays down the specific items which have to be mandatorily disclosed in the annual financial statements and also to include reporting thereon by extending the scope of the Manufacturing and Other Companies (Auditor's Report) Order by including specific certification thereof. This has the combined effect of authenticated disclosure and an independent, specific attestation, by way of a positive confirmation or otherwise, by the statutory auditors on the matters disclosed in the accounts certified by them. Any reservation, qualification or adverse remark by the auditors would require the Board of Directors to offer the fullest information and explanations under the provisions of Section 218 (3) of the Companies Act, 1956. This would also draw the attention of readers by providing the authorities or environmentally-concerned share-holders, financial institution, pollution control authority or non-governmental organisation concerned with the environment, sufficient data to initiate remedial action.

Data that cannot be quantified but indicates projections, promises or goals should be indicated in the Directors Report by way of an amendment to Section 2171A of the Company's Act.

The *Eastern Economists*, in an article on *Enviornment Accounts*, makes a pertinent point when it says that companies consider that the requirements of environment reporting have been fulfiled by printing their reports on recycled paper, with a few nice pictures of trees and flowers in their pursuit of shading their accounts green.

MEANINGFUL DISCLOSURES

To make the disclosures more appropriate or meaningful, they should be shorn of bureaucratic jargon, full of meaningless figures, with no yardstick for damage or performance assessment. The same Eastern Economist article says that publication of environmental liabilities usually includes information relating to cleaning up contamination from land, air and water which are the result of the company's activities, setting time-bound schedules to regulate air and water contamination with reduction to toxic emissions and restating its achievements vis-a-vis its targets. Figurewise-data should include publication what pollutants have been released into the air, water and soil from individual plants, and listing the extent to which they have decreased or increased during the year under report. Listing of "unwanted events" (cross-referenced with extraordinary items disclosure in accounts) like accidental spills, complaints, prosecutions and penalties, and pollution-specific compulsory shutdown, either by the government or by court order as was done by the Gujarat High Court in 1995.

FISCAL INCENTIVES

The Government of India has introduced fiscal incentives for pollution control measures in the form of accelerated depreciation on the installation of such equipment. This, by itself, does not result in corresponding reduction in pollution. The effect of this should be measured by the reporting of actual reduction and prevention in the level of pollution by presenting hard facts and figures, including the cost of installation, maintenance, and reduction on pollution levels or emissions in terms of quantified information, and introducing the concept of Polluter Pays for increases in the level of pollution and offering better incentives for the containment of pollution.

The quantified, concrete date should be in the form of more efficient use of materials, recycling, upgradation of recycling technology, use of scrap, waste, sewage and biodegradables, savings through energy-efficient methods, paper and stationery. Rather than limiting environmental reports to quantities like kilos, litres and tonnes, it would be worthwhile to express them in terms of hard cash to make it more 'auditable' and expressing them in terms of rupees to attribute 'value lost' in the same many as 'value added' is worked out. Thus, environmental audit in monetary terms will make the exercise more meaningful.

THE EUROPEAN EXPERIENCE

The European Commission has begun to demand more of such measurements by industry. American industries are required to state the details of 300 substances on their Toxic Release Inventory. The United Nations Environment Programme and the International Chamber of Commerce have launched the Public Environmental Reporting initiative as a basic framework for environmental reporting.

Arthur D. Little notes that in Europe, perhaps five per cent of the statistics on a typical environmental report emerge from continuous measurement. Another 30 per cent comes from frequent measurement, and the rest from single readings and estimates.

CONCLUSION

In India, we are not better off. We should not adopt western complacency. Rather, we should initiate an inter-disciplinary approach with the audit profession acting as the monitors and reporters by verifying, attesting and reporting on environmental disclosures to make an item a more reliable input for rewarding performers and penalising polluters.

26

*Accounting for Environmental Costs : A Hazardous Subject**

JOHN P. SURMA AND ALBERT A. VONDRA

HOW TO COPE WITH A CHANGING REGULATORY AND LEGAL FRAMEWORK

In the 1990s, corporate America began to face the environmental costs of industrial practices and consumer preferences that had seemed, only a few decades ago, to be the legitimate rights of a growing nation. This article deals with a pragmatic and critical aspect of environmental protection: corporations' accounting practices in reporting the costs of past, present and future environmental activities. Based on our 1990 survey of 125 major U.S. corporations, the article is designed to help financial and business executives better understand the environmental issues their accounting departments wrestle with daily.

According to the survey, companies that recognize their environmental responsibility by chartering board-level committees generally are viewed more favorably by shareholders. Arguably, corporations choosing this strategy will be better positioned in a future in which environmental issues occupy an increasing share of the U.S. regulatory and legislative agendas, public concern and scientific research.

Despite the growing importance of environmental issues, only 14% of survey respondents had a formal board committee. Since these issues often involve litigation, companies clearly preferred to assign oversight responsibilities to the general counsel's office. Further, only 11% had accounting policies specifically addressing environmental accounting; fewer than a third of those disclosed the policies in financial statement footnotes.

TIMING OF RECORDING CLEANUP LIABILITIES

Hazardous waste remediation (or cleanup) under the Comprehensive Environmental Response Compensation and Liability Act of 1980 (CERCLA), the Resource Conservation and Recovery Act (RCRA) or similar state or local statutes often is a company's largest cost and the most difficult to estimate. The materiality of the amounts involved, CERCLA's joint and several liability provisions and often protracted legal proceedings surrounding such cases complicate the required accounting judgments.

The most important accounting pronouncements directly addressing the recognition and measurement of environmental cleanup liabilities are Financial Accounting Standards Board Statement No. 5, *Accounting for Contingencies,* and the limited focus of Emerging Issues Task Force Issue no. 90-9, *Capitalization of Costs to Treat Environmental Contamination.* Also, FASB Interpretation no. 14, "Reasonable Estimation of the Amount of a Loss", provides guidance on recording a liability when a range of probable loss can be estimated. However, while environmental matters may receive greater FASB attention in the future, the near-term likelihood of a formal FASB standard is quite low.

The survey asked at what point respondents recorded hazardous waste remediation liabilities from prior activities. The results appear in Exhibit I. Accounting practice was mixed regarding timing of cleanup liability recognition, possibly due to the difficulty of estimating the costs' components. Some companies recognized liabilities at the sale, disposal or abandonment of facilities; others recognized them at or before completion of a remedial investigation and feasibility study (RI/FS).

Practice variations between companies reflected not only the emerging nature of environmental liabilities but also the unique nature of individual cleanup sites and environmental problems. When viewed in the context of Statement No. 5, the responses indicated *measurability* (the ability to reasonably estimate) rather than *probability* was most often the deciding factor in determining when to record a liability.

ESTIMATING REMEDIATION COSTS

Recognizing a liability for hazardous waste remediation frequently depends on being able to estimate remediation costs reasonably. Although current accounting literature provides some guidance, many other aspects must be considered.

Developing a reliable estimate requires evaluating technological, regulatory and legal factors, each of which calls for exercise of management judgment to reach a supportable accounting conclusion. The potential for third-party recoveries makes it difficult to estimate an environmental obligation's ultimate cost. The survey asked the relative importance of the variables involved in recognizing and measuring environmental remediation liabilities (see Exhibit II).

Other key factors in the recognition and measurement process included:

- Evolving remediation technology.
- Changing regulatory standards.
- Experience at similar sites.
- Existence and quality of records supporting amounts and types of waste contributed.
- Materiality of waste contributed.

EXECUTIVE SUMMARY

- U.S. Corporations have reached a critical point in accounting for environmental activities. A survey of 125 companies revealed companies were concerned about a changing legal and regulatory framework.

EXHIBIT I

Timing of Recording Liability for Cleanup of Prior-Period Waste

(Percentage of respondents)

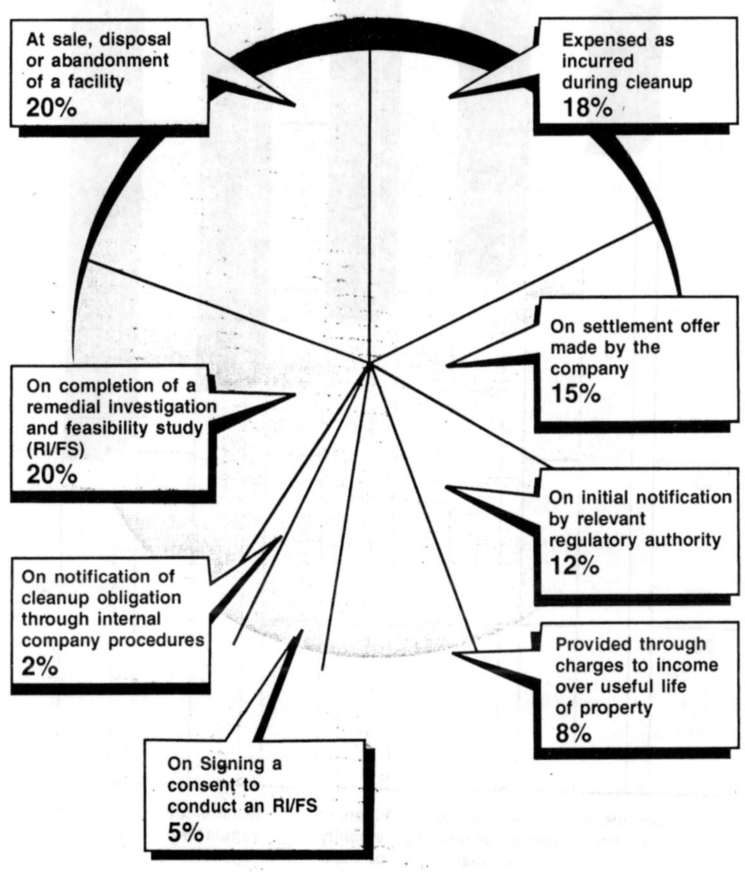

EXHIBIT 2

Importance of Variables in Estimating Liabilities

(Percentage of respondents)

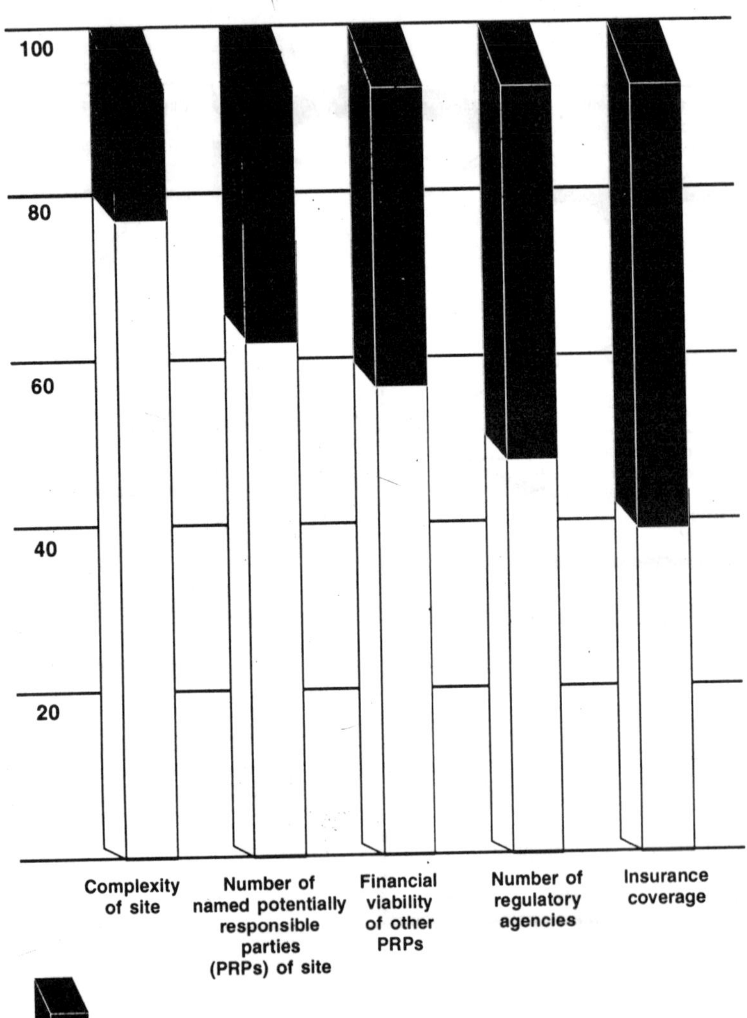

- Only 14% of Survey respondents had established a board of directors committee to oversee environmental matters. There was a clear preference to assign oversight responsibility to the general counsel's office.
- Timing of Accrual of cleanup liabilities varies. Measurement often can be difficult and practice is mixed with regard to the timing of the recording of environmental cleanup liabilities.
- Estimating the Magnitude of environmental remediation liabilities involves many variables: Site complexity and the number of potentially responsible parties were among the most significant.
- Postremediation monitoring costs at closed sites are not always considered when establishing accruals for environmental cleanup obligations. Some survey respondents cited immateriality as the reason for this practice.
- Nearly All Survey respondents included environmental costs in measuring net gain or loss on disposal of a site, using the same recognition and measurement criteria as in accounting for such matters at ongoing facilities.
- Defining Environmental costs is difficult for U.S. corporations. Most survey respondents considered legal fees, consulting fees and other soft costs to be environmental costs.

Among the factors influencing estimates, some factors favored achieving reasonable estimates while others heightened uncertainty. The survey results showed recording a loss net of expected insurance recoveries was accepted in practice, depending on individual facts and circumstances).

The Securities and Exchange Commission staff, which has taken a more active role monitoring appropriate accounting for environmental liabilities than in the past, says Interpretation no. 14 should be given careful consideration. Registrants must take reasonable steps to identify the minimum end of the loss range with a higher priority on studying sites where known problem exists.

The SEC says registrants should assess the probability of an environmental obligation and any insurance recovery independently.

In assessing the probability of an insurance recovery; companies should consider the success of similar claims and the insurer's financial viability. When an environmental contingency meets the criteria in Statement no. 5 and insurance recoveries are not probable and reasonably estimable, the liability should be recorded. The SEC staff says, it is inappropriate to reduce a probable liability with a reasonably possible insurance recovery.

POSTREMEDIATION MONITORING COSTS

Depending on the remedy and substances involved, postremediation site monitoring for 30 years or more may be required. Three different approaches to accounting for postremediation monitoring at closed sites have been adopted by survey respondents. Exhibit III, illustrates how many respondents used each approach. Frequently, these costs were not considered in establishing environmental cleanup liabilities. Since long-term monitoring of closed sites is not elective but integral to the remediation process, immateriality is the only persuasive argument for excluding such costs from remediation accruals.

Among companies reporting postremediation monitoring or other perpetual-care costs recorded as part of the cleanup obligation, there was no clear preference for recording the total future value or net present value of established costs. At present, discounting is not permitted by current accounting rules except when the future obligation is fixed and the timing of settlement is reasonably determinable.

FUTURE EXPENDITURES FOR "EXIT" COSTS

Future exit-cost expenditures can include reclamation, wastewater pond closure and the ultimate closure of production sites. Companies reporting such accruals were asked to specify the nature of events or costs being accrued and the method for estimating future liabilities. They also were asked whether they used the units-of-production or a similar method to record the estimated cost.

A majority of companies (68%) said they did *not* accrue currently for future environmental expenditures. Among those reporting accruals (32%), the following was noted:

EXHIBIT 3
Accounting for Postremediation Site Monitoring

(Percentage of respondents)

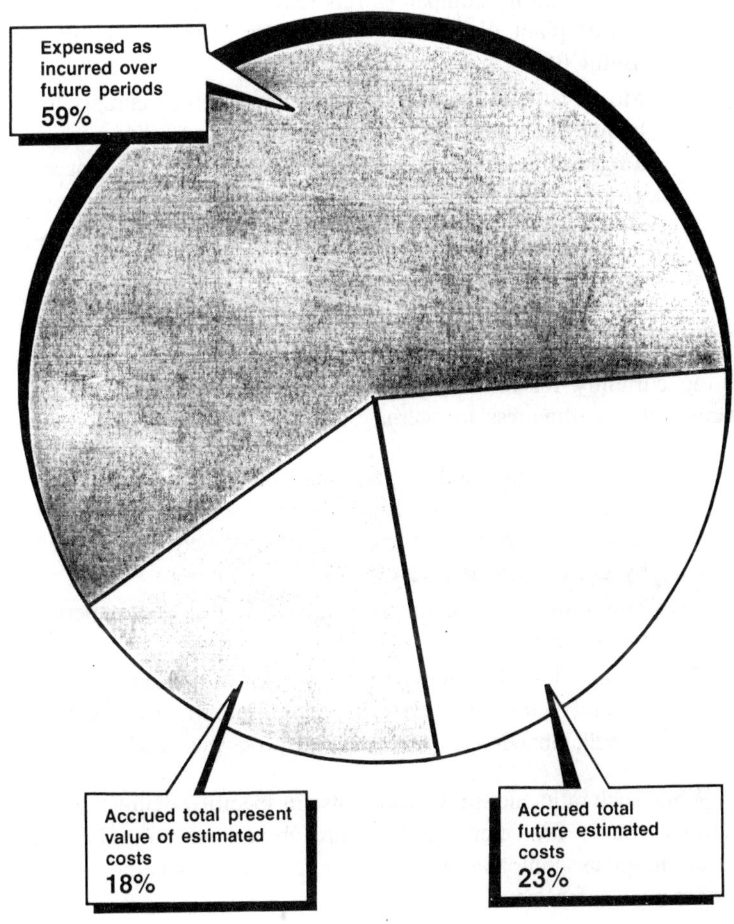

- Over 1/3 of the accruals were reported by companies with operations in the extractive industries, principally coal and oil and gas. Most of these respondents said such accruals were recorded by units of production, based on engineering estimates.
- Several utility companies reported the accrual of nuclear power plant decommissioning costs over the facility's useful life.
- Most of the remaining respondents reported accruals for known environmental incidents at operating facilities, when a reasonable estimate of the response cost could be made. These accruals addressed tank ruptures, soil and water contamination and underground storage tank leaks. Estimates were based on engineering and consulting studies as well as on negotiations with regulatory authorities.

While accounting literature can be interpreted to accommodate accruing currently for future environmental obligations, the survey reflected an unwillingness to accrue because :

- The environmental regulations and related obligations concerning site disposal were not always regarded as firm and unchangeable; some circumstances were not covered by existing regulations.
- The common expectation was most facilities would operate indefinitely.
- Companies often took the view site restoration costs in general were more properly recognized when a decision to sell, abandon or otherwise dispose was made.

While such rationales may make sense in specific circumstances, accruing currently for determinable future obligations may become more common as companies become increasingly concerned with environmental compliance costs.

SALE, ABANDONMENT AND OTHER DISPOSAL TRANSACTIONS

Environmental cleanup and other site restoration obligations need to be considered in measuring the results of sale, abandonment or other disposal transactions. Accordingly, the survey asked if respondents in such situations applied different recognition and measurement criteria for recording environmental cleanup obligations. The vast majority (92%) reported they treated these obligations the same as obligations related to ongoing operations.

Several survey respondents said they recorded environmental liabilities as an assumed liability in a purchase business combination. Environmental obligations that are contractually or legally assumed and reasonably estimable normally would be treated like any other assumed obligation in recording assets acquired and liabilities assumed. The current trend toward commissioning comprehensive environmental assessments as a part of due diligence should result in a better basis for recognizing environmental obligations assumed in business combinations.

THE BURDEN OF COMPLIANCE

The burden of complying with evolving environmental regulations is increasing for many U.S. corporations. While such costs do not pose significant accounting problems, how broadly can environmental costs be defined? The survey listed 20 distinct categories of administrative costs, ranging from internal payrolls to various outside consulting fees, and asked respondents to identify which would be treated as environmental costs for disclosure or accrual purposes, or both.

For accrual purposes, the majority of companies would include outside legal fees, consulting fees and other soft costs related directly to a specific environmental problem. They would exclude internal payroll costs except those incurred for remediation. In addition, a majority of respondents believed more generalized activities as well as the ongoing cost of waste reduction programs were period-operating costs and not environmental costs for accrual purposes.

BROAD GUIDELINES

The survey results provided new insight on an increasingly important accounting challenge. The following broad guidelines should be considered in evaluating environmental exposures.

Cash Basis Accounting

The pay-as-you-go approach for environmental cleanup obligations is inconsistent with accrual accounting and Statement no. 5. It should be questioned unless the amounts involved clearly are immaterial.

Amortization of Costs

No conceptual or practical support exists for amortizing environmental cleanup costs related to past activities through periodic charges to future income.

Expense Recognition

For environmental cleanup costs under CERCLA, RCRA or similar statutes, a strong presumption exists for recognition no later than completion of the RI/FS or similar phase. Other variables, such as participant share and third-party recovery, also should be assessed to determine if there is a basis for delayed recognition or recording less than the gross estimated liability.

Net versus Gross

Recording an environmental liability net of expected recoveries is often accepted in practice, depending on recovery likelihood.

Plant Shutdowns and other Disposals

Environmental cleanup and other site restoration obligations should be considered in measuring sale, abandonment or other disposal transaction results. Most companies include environmental costs in measuring a net gain or loss on site disposal, using the same recognition criteria as in accounting for ongoing facilities.

Business Combination Liabilities

To the extent reasonably estimable environmental obligations are assumed in a purchase business combination, they should be treated like any other assumed obligation in recording assets acquired and liabilities assumed. Increased use of environmental assessments in negotiating business transactions may lead to more frequent recognition of environmental obligations.

Present Value Approach

Use of present value techniques, other than those called value under Accounting Principles Board Opinion no. 16, *Business Combinations,* and no. 21, *Interest on Receivables and Payables,* is currently the subject of a FASB project. Until additional guidance is provided, accruing remediation costs at present value should be approached cautiously and restricted to times when future cash flows are highly predictable.

Recognizing a liability for hazardous material remediation depends on developing a reliable estimate of the liability—evaluating numerous technological, regulatory and legal variables. A complex regulatory structure has evolved in the United States that has an impact on virtually all enterprises. Financial and business executives must learn to manage, control and, most important, predict uncertainties to ensure their enterprises' future competitiveness. As the spotlight on environmental issues becomes more focused, cleanup technology and equipment improve and estimating cleanup costs becomes easier, earlier recognition of liabilities in financial statements may result, bringing greater structure and consistency to environmental accounting.

27

Pollution Control through Social Cost Conversion

FLOYD A. BEAMS AND PAUL E. FERTIG

The role of accounting in our current ecological crisis is not passive. Accounting provides information upon which decisions are made—decisions that result in economic and social actions. If the resulting activities disrupt the environment then accounting is, at least in part, accountable for that disruption.

Accounting as an organized profession has the responsibility to transcend the internal viewpoint of a private firm and to develop information which portrays private firm's role in and contribution to society.[1] Accounting information should lead to decisions that result in the efficient utilization of resources, the conservation of the environment and the equitable allocation of business income. Accounting can assume an active role in controlling pollution by working toward the concept of production costs suggested by President Nixon: "To the extent possible, the price of goods should be made

Reprinted with permission from the *Journal of Accountancy*, copyright 1971 by the American Institute of Certified Public Accountants, Inc. Opinions of the authors are their own and do not necessity reflect policies of the AICPA (November, 1971, pp. 37-42).

to include the cost of producing and disposing of them without damaging the environment."[2]

This article seeks to show that reluctance of our industrial firms to accept responsibility for damaging the environment is likely to be a major obstacle in our attempts to control pollution. This statement from an executive of "a major U.S. oil and chemical company" may be typical: "Many of us already have the money to spend on cleaning up the dangerous level of pollution, but we are not going to spend it until all of us are forced to do the same."[3] Similarly, accountants have not taken the lead in accepting pollution cost as a private cost for accounting purposes. For example, a recent article entitled "Production Costing in Open Pit Mining" does not mention the possibility that the cost associated with defacing the earth's surface should be considered a production cost.[4] Such omission may not be realistic in view of increasing social pressure on these firms to accept responsibility for the damage caused by their production activities and efforts of some firms to correct it.[5] As we shall see, firms and industries differ greatly with respect to the amount of pollution they create, as well as in the efforts they expend to neutralize it. These differences severely impair the comparability of income reports. Under current reporting practices, those firms which appear to be efficient may be very inefficient. Those firms which have adequate earnings records may be economic failures. Those companies receiving the savings of investors may be the least deserving of additional resources.

HOW ACCOUNTING CONTRIBUTES TO POLLUTION ?

The dominant motivation for business activities is usually considered to be the profit motive. Profit maximization requires, among other things, the minimization of costs for any given level of activity. This is economic theory. But in accounting for a firm the theory regresses to the proposition that the minimization of private costs is a necessary condition of private profit maximization. Private costs are those measurable pecuniary costs of resources used or destroyed in conducting business activity that are borne by the firm. Not included in the profit maximization formula are those additional costs of resources used or destroyed in conducting business activity that are borne by others or by society. These are the social costs of business enterprise.[6] They are costs in the sense that they represent resource values destroyed by the firm and paid for by others. Typical of such resource destruction is pollution of air, contamination of the streams and deterioration of residential and industrial sites.

Basically a firm which is motivated by a desire to maximize private profits has no incentive to conserve resources which are free goods from its internal point of view. Thus, the expenditures which a firm makes to conserve environmental resources generally conflict with the private profit objective. Not only is there a failure to conserve environmental resources, but the combination of the private profit objective and current accounting practices may lead to excessive use and misuse of air, water and other resources. This occurs when private costs are minimized by wasting environmental resources. Examples of this situation are the dumping of untreated sewage and unfiltered air into the environment.

Such dumping increases social costs while holding private costs to a minimum. The result is a type of income redistribution from citizens and taxpayers to the customers and investors of the polluting firm. The process of minimizing private costs by increasing social costs is also objectionable because the production and distribution activities of a polluter are likely to be inefficient. This will be the case when the total private and social costs of output exceed the private costs which would have to be incurred to produce and distribute the same output without damaging the environment.

These problems which arise from excluding social costs from income measurements are not new. Many accountants have written about the problems and the American Accounting Association has organized committees to conduct research in the areas of Socio-Economic Accounting (1969-1970) and Measures of Effectiveness for Social Programs (1970-1971). The current concern for pollution control and environmental replenishment has created a new urgency for the profession to accept an active role in regard to social costs. On the one hand, accounting is being criticized for contributing to the decay of our environment and, on the other hand, new organizations are being formed to provide information relating to the social responsibility of corporations.

In response to the demand for more information on private enterprise than is provided in typical financial statements, a new organization called the Council on Economic Priorities (CEP) was created. This organization investigates corporations and issues reports on its findings. The purpose of the reports is explained by Alice Tepper, director of the organization: "We would simply like to see social responsibility become, like profits and earnings figures, a standard by which corporate practices are evaluated and exposed to the inivesting public."[7]

Embedded in this statement are the implications that investors may misdirect their savings by over-looking the social responsibility aspect of corporate operations and that public pressure may be applied for firms to acknowledge social responsibilities. As a stopgap measure, such reports on social responsibility may be tolerable. But in fairness to all parties concerned, the information should be generated according to data accumulation, reporting and attestation principles. Without principles the confidence of the American public cannot be acquired.

There is a tremendous demand for this type of information on social responsibility from widely diverse groups (investors, lawmakers, politicians, economists)—and it is not available. Organizations like the CEP are responding to the demand. Some organizations may try to make impartial, unbiased, complete and accurate reports to the best of their ability—whatever it may be. Other organizations may themselves simply be responding to a profit motivation. These organizations are without standards, without discipline and lacking the professionalism of accountancy with its guardian, the AICPA.

Some perspective as to the nature and extent of current criticism is provided by the following excerpt in which Professor Quigley of the department of history at Georgetown University indicts the accounting profession for its role in the decay of our environment.

> The "firm" was an innovation in bookkeeping techniques, just as the "corporation" was a legal gimmick. Both were man-made and both are imaginary; yet together with the industrial revolution they have made it possible, even likely, that we have already passed the point of no return in environmental pollution.
> Establishment of the "firm" was a bookkeeping decision that in calculating profits by subtracting "costs" from "income," "economic costs" would be included but "social costs" would not be counted.[8]

This indictment of accounting was not a passing fancy in Quigley's article; it was his theme. He further explains that Americans

> . . . use falsified accounting techniques and mistaken taxation methods, not only to encourage this process [environmental decay], but to conceal from themselves what is really happening.[9] The ultimate falsehood of our accounting is to be found in official and semiofficial statistics on the American "standard of living."[10] Moreover, this whole system of false reporting on the condition of America is solidly sustained by the tax system since the upkeep and maintenance of the most destructive earth-mover is tax deductible.[11]

The remedies for our ecological crisis as listed by Quigley were (1) change the tax system and fiscal policy, (2) revise corporation law and (3) create a system of social accounts.[12]

Although few accountants would be willing to accept Quigley's arguments (to say nothing of his language) or the degree of responsibility he seems to imply that we should accept, the basic message is unmistakable and cannot be ignored. Many segments of our society are at fault and accountants are not beyond criticism.

NON-COMPARABLE FINANCIAL REPORTS

The problem of comparability in financial reporting practices has received considerable attention from the accounting profession over the years. Although many of the problems have never been solved, considerable progress toward more comparable reports has been made. Now the problem of omparability arises again in connection with social costs and environmental pollution. Some companies within an industry are controlling pollution by making current outlays while others are seemingly ignoring the pollution problem. The operations of some industries are more destructive to the environment than the operations of others. Programs of abatement and costs of abatement will also vary by company, by industry and by location. The probable result of these differences is that, other things being equal, the worst polluters will appear to be the most successful and they will likely receive additional resources from the investing public.

An example of these variations comes from a recent government report which estimates that ". . . over half the volume of wastes discharged to water comes from four major groups of industries—

paper manufacturing, petroleum refining, organic chemicals manufacturing and blast furnaces and basic steel production."[13] The Council on Economic Priorities mentioned above recently issued a report on the paper industry that notes that "the paper industry has been generally slow to install anti-pollution devices and processes, despite their ready availability. Owens-Illinois and Weyerhaeuser are important exceptions; both companies clean up most of their plants' effluents. Less than half of the 131 mills surveyed have satisfactory air pollution controls; many dump raw wastes into U.S. water-ways."[14]

The significance of those social costs which are currently being omitted from the profit calculations of most firms is suggested by the following estimates of Professor Goldman. Goldman estimates that the annual operating costs of pollution control will run 1% to 2% of GNP and 4% to 7% of the value of industrial, agricultural, mining and transportation output. He compares these estimates to the experience of industrial firms which sometimes allocate as much as 10% of their expenditures to pollution control.[15]

Another source estimates that air pollution control will cost $1.94 per $100 of sales for the average iron and steel mills, $2.89 for iron foundries, $3.92 for nonferrous metal plants, $.21 for grain mills and $.95 for cement plants.[16]

Still another source has estimated that industrial spending on pollution control during 1971 will range from 3% to 5% of a total capital investment of about $70 billion—or as much as $3.5 billion.[17]

Comparability of financial reports with respect to pollution cost cannot be achieved merely by insisting on complete disclosure of the voluntary efforts of firms to neutralize pollution and correct the damage. What is needed is a comparison among firms of the damage created by production activities. This comparability can be achieved only through the application of a full-fledged system of accrual accounting.

ACCRUAL ACCOUNTING FOR POLLUTION

The essence of the application of accrual accounting to costs of pollution is based on the proposition that the costs of neutralizing the damage to the environment are costs of production. This means, of course, that costs incurred in connection with current manufacturing activities should be treated as current product costs, and costs incurred to neutralize future pollution should be capitalized and allocated to future manufacturing activities. Costs associated with repairing environmental damage resulting from activities of prior periods should be accounted for as a correction of prior periods' income. Identification of costs of past, current and future activities will often be difficult, but the effort must be made.

The financial position effect of recording these costs may be to change property values as industrial sites are deteriorated and re-established or it may be to recognize liabilities which have neither asset implication nor legal status. These alternate forms are developed below.

Resource Impairment

The value of an industrial site is dependent upon adequate air and water resources and upon efficient means of waste disposal. When these environmental qualities which give a site its value are impaired, the costs of such impairment must be reflected in the financial reports. If the environment is contaminated by a firm's own production activities, the cost of resources destroyed should be associated with the production activity which gave rise to the site deterioration. Site deterioration due to the actions of other firms or to changes in regulations which reduce the efficiency by which industrial wastes (gaseous, liquid or solid) can be discarded may be considered period costs or losses.

Accounting for site deterioration is similar to depreciation accounting with one important exception. Depreciation cannot be avoided. By contrast it is possible to maintain the environmental quality of an industrial site or even to improve the environmental quality of a site which has been allowed to deteriorate.

Thus the application of accrual accounting suggests that outlays for pollution control that result in a maintenance of existing environmental conditions are current expenses. Outlays to re-establish the environmental quality of a deteriorated site are capital expenditures, provided, of course, that the earlier deterioration was charged to expense or to production. Pollution control outlays to provide for future site maintenance should be capitalized and spread over the period of expected benefit.

Accelerated obsolescence is another area in which accrual accounting applications must be kept modern. Depreciation rates for plants and equipment which do not meet existing pollution control standards may not reflect the obsolescence factor. Consider the following observation of a steel company executive:

> If we put in a unit in 1957 or 1958 to meet the code at that time, we have by no means written it off or gotten our money out of it. Now we're hit with a new standard and we're faced with quite a writeoff, because in most cases it isn't a matter of just adding on; it might mean a whole new facility.[18]

Other examples of actual or potential deterioration of property values resulting from new standards of industrial waste disposal can be found throughout current news. Consider the accounting ramifications of the following regional news items, for example:

The State Water Control Board of the Commonwealth of Virginia has set July 1972 as the deadline for American Cyanamid Company's Piney River plant to cease polluting the Piney River. The high cost of the new controls which would be necessary to halt the pollution led to the announced closing of the plant and the possibility of throwing 325 people out of work.[19]

The Saltville, Virginia, soda ash plant of Olin Mathieson Corporation is scheduled for closing by 1972. The history of this plant dates back to 1893 and the reason for closing is simply that the plant cannot meet the prevailing water quality, control standards. About 650 people will be out of jobs if the anticipated shutdown occurs.[20]

Strip mining practices will be banned if the Hechler Bill which proposes to outlaw strip mining becomes a federal law. Strip mining and related operations are estimated to employ about 5% of the labor force in West Virginia.[21]

Liability Recognition

In addition to the above proposal that resource impairment which results from environmental pollution be recognized and reported on an accrual basis, we propose that expected future outlays for environmental damages which result from past and current production activities be estimated and reported on a current basis. Admittedly, the measurement problems are difficult, perhaps even insurmountable at the present time, but the difficulty makes the problem no less important.

Liability accounting for environmental pollution may take any of several forms. The legal liabilities which result from a firm's violation of existing laws must be reported. Contingent liabilities from probable actions where firms are in violation should also be reported. In addition, firms should accrue liabilities for those expected future outlays which will not create asset values for the firm.

Just as lessees accrue liabilities for restoration of leased property at the termination of their leases, mining companies should accrue liabilities for eventual land restoration during the stripping operations. Similar liabilities should be accrued for the manufacturing companies that are slowly assuming responsibility for cleaning up segments of the public domain. That is, when industrial companies clean up beaches, lakes, rivers and so on, they are not creating property values for the firm. Accrual accounting would seem to require charges to income in the period of contamination rather than in the period of outlay.

There is some evidence that firms would prefer to use cash-basis accounting for pollution control outlays. The following query and answer from an interview between a *U.S. News* reporter and Edwin H. Gott, chairman of U.S. Steel Corporation, illustrates this point.

> Q. Is there some way the cost of adding pollution abatement equipment could be passed on to the public other than by price increses?
>
> A. The problem now is that pollution control equipment is classified as a capital improvement. You write the costs off as part of the normal depreciation process. I think the more practical way would be to consider installation of this equipment as an operating expense, part of the current cost of doing business. The installation of the equipment is not done for profit motives, but strictly to comply with regulations.[22]

THE MEASUREMENT PROBLEM

Unfortunately, concerted social action for pollution control has been postponed until the point at which many recognize it as a national emergency and crisis. Time is not available to conduct the research necessary to establish verifiable methods for measuring the costs of pollution at the level of the individual firm. Instead, it appears that pollution costs to the firm in many industries are to be determined by law and regulation. Firms will incur costs, not necessarily to neutralilze the environmental damage they cause, but to maintain the community at some legal standard established by government agency. The enforcement of legal standards could take the form of assessing effluent charges against firms for industrial wastes which they dump into the waterways or the atmosphere. Monitoring devices to detect and measure different types of effluents are available and are being installed in many areas of the country. Effluent rates could be set arbitrarily high to encourage firms to avoid the charges by developing their own programs of abatement. Alternatively, the rates could be set so as to finance regional projects of pollution control.

Perhaps this method of pollution control is unavoidable, but it means that many of the problems of measuring the cost to the firm of controlling pollution will be a matter of law, rather than of technology or economics. This will not be universally true, of course, and the opportunities for voluntary assumption of pollution costs by individual firms will remain in many industries. Therefore, research into the pollution costs to the firm is crucial, and the results can also be used to identify and correct inequities in the laws as they occur. In the meantime, the accountants' task in the area of measurement is to adapt to changing regulations and costs structures which affect nearly all phases of the reporting system.

RECOMMENDATIONS

In view of the foregoing discussions, we propose that published financial reports disclose information on the responsibility for pollution whether imposed by law or assumed voluntarily by firms. Our proposal is based on the following premises.

First, we have a national commitment toward controlling and reducing environmental pollution. Firms will pay for and pass on to the consumer the cost of environmental resources which are destroyed in the production and distribution processes. Thus, the decisions firms must make are concerned with which pollution control programs to undertake and when to undertake them, within the limits of the law.

Second, in the absence of adequate measurement methods, firms will generally account for pollution outlays as expenses. As indicated above, cash-basis accounting will make the income statements of delinquent firms appear more favorable than statements of firms which are voluntarily assuming responsibility for pollution control. Accounting need to apply accrual accounting procedures to pollution costs in order to obtain comparability in published reports. The accounting reports should also provide additional disclosures that will give the reader some basis for assessing the responsibility that firms or industries should assume. The demand for information on the social responsibility of firms will be met by others if accounting does not rise to the challenge of supplying it.

Accounting can and should provide disclosures on pollution costs so that financial statement users are informed of company efforts at abating pollution and of the possible lack of comparability in financial statements of different firms. The quantitative data presented in financial reports should meet accounting tests for data accumulation and reporting. All disclosures should meet the tests for attestation.

Our proposal consists of two parts—a verbal description and a quantitative reporting requirement. The verbal description would consist of a dual disclosure in general terms of firm and industry position with respect to pollution control problems. An introductory statement should appear on the financial statements to the effect that a valid comparison of the financial statements of different companies requires consideration of pollution control programs. Keyed to this statement should be: (1) a standard industry footnote which identifies the major pollution control problems within the industry, the goals of the industry in abating pollution, the control standards which have been imposed and the deadlines for compliance with existing standards (this industry note could be prepared by the research staff of the AICPA and supplied to auditors on a quarterly or semiannual basis) and (2) a firm disclosure which relates to the industry disclosure and compares firm and industry pollution problems, regulations, deadlines, goals and programs of abatement.

The second part of the proposal is to require separate disclosure in financial statements of pollution control expenses (income statement), pollution control outlays (funds statement) and pollution control resources (balance sheet). This information will give an investor a basis for determining how one firm's pollution control program compares with the programs of other firms. The current trend toward publicizing dollar expenditures for pollution control is misleading because equal dollar amounts have different implications in different firms and industries. Again the AICPA could serve the profession by providing suggestions for standardized terminology and guidelines for classification of pollution control items.

The recommended disclosures would give investors and the public an opportunity to assess a management's real commitment toward pollution control. They could also provide a stimulus for management to assume an earlier and voluntary responsibility for damages to the environment due to the firm's activities. Such disclosures could lead to an acceleration of the process of transferring social costs into the private cost framework. Procrastination of some firms in undertaking pollution control programs prevents upward adjustments of product prices and puts the socially conscious firm at a competitive disadvantage in terms of EPS and stock prices. The recommended disclosures could offset the current EPS advantage of the procrastinators by encouraging investors to rely on the additional information to predict long-run EPS for all firms.

Notes and References

1. The reader may wish to contrast this viewpoint with Professor Paton's statement that "the notion that the goal of the professional accountant is public or social service is nonsense. His function is to provide the best possible service to his specific clients, the people who pay for his efforts." William A. Paton, "Earmarks of a Profession—and the APB" (JofA, Jan. 71, p. 41).
2. 1970 State of the Union Message.
3. "Stepped-up War on Pollution," *U.S. News and World Report,* January 11, 1971, p. 20.
4. Robert H. Davis, "Production Costing in Open Pit Mining," *Management Accounting*, January 1971, p. 39.
5. "Hiding the Scars," *Wall Street Journal*, May 24, 1971, p. 1.
6. Observe that the concept of social costs of business, enterprise is merely a part of the broader concept, social costs. Floyd A. Beams, Ph.D., is associate professor of accounting at Virginia Polytechnic Institute and State University in Blacksburg, Virginia. Dr. Beams, a member of the American Accounting Association and the National Association of Accountants, has had articles published in several professional periodicals. He received his Ph.D. degree from the University of Illinois in 1968. Paul E. Fertig, CPA, Ph.D. is professor of accounting at The Ohio State University. Dr. Fertig is a member of the American Institute, the American Accounting Association, the Ohio Society of CPAs, the National Association of Accountants, the Financial Analysts Federation and the Financial Executives Institute. He coauthored, with Donald Istvan and Homer J. Mottice, Using Accounting Information, the second edition of which was published this year by Harcourt, Brace Jovanovich,

Inc. Dr. Fertig was the AICPA Distinguished Professor of Accounting at Virginia Tech. for 1970-71.

7. "Report on Paper," *Time*, December 28, 1970, p. 41.
8. Carroll Quigley, "Our Ecological Crisis," *Current History*, Vol. 59, No. 347, July 1970, p. 9.
9. *Ibid.*, p. 11.
10. *Ibid.*, p. 12.
11. *Idem.*
12. *Idem.*
13. U.S. Department of Interior, Federal Water Quality Administration, *Clean Water for the 1970's: A Status Report*, June 1970, p. 5.
14. "Report on Paper," *Time*, December 28, 1970, p. 41.
15. Marshall I. Goldman, "The Costs of Fighting Pollution," *Current History*, Vol. 59, No. 348, August 1970, p. 81.
16. "Pollution Price Tag: 71 Billion Dollars," *U.S. News and World Report*, August 17, 1970, p. 41.
17. "Stepped-up War on Pollution," *U.S. News and World Report*, January 11, 1971, p. 21.
18. "How Private Industry Combats Pollution," *U.S. News and World Report*, February 15, 1971, p. 46.
19. Summarized from "Holton Joins Fight to Save Nelson Plant," *The Roanoke Times*, December 24, 1970, p. 7.
20. Summarized from "The Price of Ecology: A Small Town Faces Challenge of Industry Loss," *The Roanoke Times*, December 6, 1970, p. B-1. Also see "End of a Company Town," *Life*, March 26, 1971, pp. 37-45.
21. Summarized from "Strip Miners Said Racing the Clock," *The Roanoke Times*, April 13, 1971, p. 7.
22. "Foreign Threats to a Basic Industry," *U.S. News and World Report*, October 26, 1970, p. 67.

28

Environmental Audit

UMESH HOLANI

The discipline of Environmental Auditing (E.A.) has experienced significant growth and evolution over this decade through the world. Apparently, it has got several discrete dimensions. First of all overall country approaches to environmental issues. Secondly, the issues bearing on environment in regard to different sectors of the economy. Lastly, the company approaches to various environmental issues. Apparently, E.A. should embrace not only all the constituents of what goes on by the brand name but also the impacts generated thereupon and elimination or minimization of any deleterious effects. The concern so far expressed has been with regard to disruption of the eco-systems.

Over the past few years industry has increasingly come to realize that sound environmental management can be equated with good management. Morever, better environmental management ensures resource saving and hence helps cut down production cost. Recycle and reuse of wastes have led to cost saving in many chemical process industries. Several industries have adopted cleaner technologies that generate less waste and make production more profitable. Industry can, therefore, clearly benefit from a critical self examintion of the processes and technologies it employ to see in which areas there is scope for improvement and foresee the potential problem areas, particularly pollution and human health. E.A. is clearly part of that self-analysis.

E.A. aims at examining the positive and negative effects of the activities of an enterprise on environment. According to the International Chamber of Commerce "E.A. as a basic management tool comprising a systematic, documented, periodic and objective" evaluation of how well environmental organization, management and equipment are performing with the aim to safeguard the environment by:

— Facilitating management control of environmental practices,
— Assessing compliance with company policies, which would include meeting regulatory requirements.

E.A. is a series of activities undertaken on the initiative of an organisations management to evaluate environmental performance. It provides an indepth review of the company processes any progress in realizing long term strategic goals.

There is absence of accepted E.A. practices with so many corporations approaching E.A. from different directions. E.A. has helped corporations to assess their performance, correct deficincies and reduce risk to the health and improving safety. Following are the various factors which encourage E.A.:

— The increasing pressure of various regulations on the company and the related cost.
— Corporate environment, health and safety pressures.
— The threat of legal and financial liabilities under various Acts.
— Increasing public and media awareness towards corporate matters.
— Senior mangers are looking for assurance that their corporate environmental performance is adequate.
— The E.A. has gained momentum because of awareness of workers, staff and Unions governing them insisted upon employers for ensuring improved safety.

Following are the objectives of E.A. :

— To identify and document the environmental compliance status.
— To provide assurance to top management.

— To increase the overall level of environmental awareness.
— To protect the company from statutory liabilities.
— To assist facility management in improving environmental performance.
— To develop overall environmental control systems.
— To develop a basis for utilising environmental resources.

Following are the benefits of E.A.:

— To ensure cost-effectivre compliance with statutes, regulations and company policies.
— To saving of cost from improved practices.
— To prepare a data-base relating production to pollution potential of all facilities.
— To develop the social reputation of the organisation.
— To develop confidence of customers and society towards output.
— To develop labour cooperation and confidence.
— To develop overall awareness, identifying problems and areas of risk.
— To encourage the use of low waste technologies and prudent use of resources and to identify potential hazards and risks.

It is unfortunate that there is a lack of awareness about pollution and environmental hazards among the general public in our country. After the event of Bhopal Tragedy in 1984, many companies have come to realize the significance of environmental management. We have a plethora of enactments which will compel the industries for promoting the need of environmental protection. There are more than 200 enactments that have been enforced in the past few decades. To mention a few:

—The Factories Act, 1948
—Water (Prevention and Control of Pollution) Act, 1974
—Water (Prevention and Control of Pollution) Cess Act, 1981
—Forest Conservation Act, 1980
—Air (Prevention and Control of Pollution) Act, 1981
—Environment Protection Act, 1986
—Hazardous Waste (Management and Handling) Rules 1989

—Public Liability Insurance Act, 1991

—Motor Vehicle Act, 1939.

The Environmental (Protection) Act, 1986 under the rule 14 requires an industry to submit annual Environmental Statement by the 30th Sept. every year from 1993 onwards to the relevent State Pollution Control Board. Rule 14 is applicable to any industry which possesses or requires consent/authorization under Water (Prevention and Control of Pollution) Act, 1974, Air (Prevention and Control of Pollution) Act, 1981, and/or Hazardous Waste (Management and Handling) Rules, 1989.

The following environmental issues are predicted to be significant over the next decade. The issues will remain the same, however the types of debates may be changed. The issues are:

(A) AIR Issues

— Global climatic changes

— Stratospheric ozone depletion

— Urban air pollution

— Air toxics

— Acid deposition

— Indoor air pollution

(B) Water Issues

— Estuaries, Coastal and Ocean Pollution

— Non point source water pollution

— Ground water contamination

— Insufficiency of water resources

(C) Waste Issues

— Inadequate disposal capacity for solid and hazardous waste.

— Long term disposal concerns for nuclear waste.

— Deep-sea disposal of wastes.

— High volume, low toxicity industrial wastes.

— Clean up and remediation of abandoned hazardous waste sites.

(D) Biodiversity Issues

— Endangered Species

— Wetlands loss
— Deforestation and loss of old growth forests.

(E) Land Use Issues

— Loss of open space.
— Competing use of public lands.

(F) Product Issues

— Environmental impacts of consumer and industrial product and product packaging.

(G) Other Issues

— Exposure to electro-magnetic fields.
— Environmental impacts of bio-technology.
— Workers exposure to toxic substances.
— Exposure to naturally occurring radiation.

Following are the important steps for proper environmental management:

1. A regular and periodic review and appraisal regarding the prevailing conditions and the improvements effected vis-a-vis the investments made and planned for.
2. Implementation of the plants with an eye to not only economic development of funds but also to selectivity of the contents and features of each of them.
3. Various companies, Institutions, Municipal Corporations, State Government and Central Government are trying their best to maintain environment intact. Careful attention is therefore, necessary so that efforts at the national and corporate level do not run counter to each other.
4. Creation of national parks for preservation of flora and fauna is an effort to protection of the environment from any kind of degradation.
5. Development of the accounting formats for keeping records of all the efforts alongwith measurable and national benefits and disbenefits.

The Indian tradition of conservation can be traced back to

prehistoric times. The theme of conservation pervades Indian culture, art and all its religions. In the modern time industrialization has posed a threat to the quality of water through effluents discharged from industries. The accelerated industrial tempo has also given rise to problems of air pollution. Metropolitan cities are plagued with automobile pollution as well. The Ministry of Environment and Forests has been entrusted with the responsibility to harmonize environmental protection with development. A number of initiatives have been taken in the direction of environmental protection over the last few years. According to the Indian Institute of Ecology and Environment, out of the nearly 4,000 polluting industries more than half have already installed pollution control equipments. The last decade has also seen the rise of public involvement in concern for the environment in India. Silent Valley, the Doon. Valley limestone quarries, the Chipko Movement, protest against the Narmada and Tehri marked a beginning. Efforts will have to be made to create greater awareness in this regard.

From the accounting, auditing and reporting aspects of environmental protection, some companies in India do make policy statements in their annual financial statements. They include Ranbaxy Laboratories Ltd. Hindustan Lever Ltd., Shaw Wallace Group etc. Environmental auditing is yet to take shape in India. E.A. is a strong management tool and should be used by industry for their own self-assessment. It also exhort the industry to resort to transparent environmental auditing which ought to be subjected to monitoring and public scrutiny. The neglect of the responsibility of industry, could precipitate public action since there was an increasing awareness among the public to have a clean environment. It is heartening to note that industry is responding to the governmental guidelines on environment. There is, however no seperate accounting standard to the effect.

References

The Factories Act, 1948.
Water (Prevention and Control of Pollution) Act, 1974.
Forest Conservation Act, 1980.
Air (Prevention and Control of Pollution) Act, 1981.
Environment Protection Act, 1986.
Public Liabilities Insurance Act, 1991.
"Environmental Auditing" Technical Report, Series No. 2, United Nations

Environment Program, Industry and Environment Office, 1990.

"Outline of Environmental Auditing", Ministry of Environment, Govt. of India, Discussion Paper, 1992.

Buckley, Jaff Perspectives in Environmental Management, Springer Verlag, Berlin (1991).

Buckley, R.C., "Environment Planning Techniques," SADME, Adelaide, 1987.

Bala Krishnamoorthy, "Environmental Trends, Issues and Key Questions" *The Management Accountant*, Aug. 1995

Forber S., Castanza R., "The Economic Value of Wetlands Systems, *Journal of Environmental Management*, 1987, 24, 41-51.

M. Ravi Sundar, "Environment Accounting", *The Management Accountant*, Sept. 1997.

N. Rajaraman, "Environment Audit," *The Management Accountant*, Sept. 1997.

P. Chattopadhyay, 'Environment Accounts', *The Management Accountant*, June 1994.

Surojit Bose and Dr. Alka Parikh, "*The Environment Audit: Holy Grail or Essential Management Tool*" *The Management Accountant*, June 94.

Sripati Rangananda, 'Environment Audit', *The Management Accountant*, June 1994.

S. Venu, "Accounting for the Environment: Corporate Perspective", *The Management Accountant*, Sept. 1997.

S. Rathore, "International Accounting," Prentice Hall, 1996.

29

An Introduction to the Concept of Energy Audit

M. SELVAM AND KODEEAWARA RAMANATHAN

INTRODUCTION

Energy short falls and the resultant economic stagnation are direct challenges to our India polity. The role of energy in our economic life is very vital and the wheel of economic progress can not move without the essential input, viz., Energy. It is a well known fact that the identification of new alternative energy sources including renewable energy resources normally takes many years and such sources may not be adequate to meet the current needs of energy in India. Hence the need for energy conservation.

The efficient energy consumption/energy conservation offers a practical means of achieving our national developments/goals. Energy conservation is not an arbitary reduction of energy consumption. Energy process has an element of avoidable and unavoidable losses and energy spent rationally. Energy organisation aims at reducing energy consumption through all possible ways like:

— elimination/reduction of wastage of energy.

— use of energy more efficiently by improved methods and process.

It is in this context that energy audit gains significance. Energy audit is an effective tool to achieve maximum energy conservation/ minimum energy consumption in an industry. Energy audit is an integrated approach covering the entire energy processs and it should be carried out without any intermission. Energy audit is a fundamental part of any management programme of any organisation which wishes to control its energy costs.

ENERGY AUDIT

The energy audit is key to a systematic approach for decision-making/planning in the energy management program. It results in identification of energy conservation and attempts to balance the total energy inputs with its use. The energy audit serves to identify all the energy streams in a facility/system which quantifies energy use. It is a positive approach which adds to the financial performance of a company. Financial audit is concerned with the profit, loss and income and expenditure of an organisation whereas the energy audit is about production and uses of an energy. Energy audit is a tool for analysing and controlling of any system where energy is used.

Energy audit helps in energy cost optimisation, energy pollution control, energy safety measures and suggest appropriate methods to improve the operating practices without affecting the quality of the system. It may be instrumental in coping the situations and variations in energy cost, availability and reliability of energy supply for sustainable development. It helps in taking decisions on appropriate energy mix, decisions on improved energy conservation equipments and ad-hoc technology. Energy audit techniques are better than the piecemeal solutions which incorporate measures without adopting a total system approach including gearing-up organisational structure and infrastructural requirements. It would give a positive orientation to energy cost reduction, preventive maintenance and quality control. Energy audit shows to the management the realistic scope for fuel savings and helps in allocation of energy resources. The effective energy audit helps in the following ways.

— to monitor the energy consumption.

— to measure the thermal and overall energy efficiency.

- — to reduce and pinpoint the wastage,
- — to workout heat balance and mass balance and time consumption.
- — to reduce the overall cost and increase the profitability.
- — to replace energetically inefficient equipments.
- — to involve the substitution of fuels.
- — to determine the effectiveness of the present energy utilisation.

Scope of Energy Audit

The scope of an energy audit varies in accordance with the facilities being audited and one extreme form of facility is a manufacturing operation where lighting, ventilation and air-conditioning are the major source of consumption of energy. The other extreme is integrated process units like refineries and petrochemical plants where cascading of energy and complex energy balances are involved. Energy audit is a pre-requisite to an energy management programme. The historic audit provides an overall picture of energy cosumption and production, so that, specific energy consumption and costs could be estimated. It is possible to workout an overall index.

The different Scopes of Energy Audit are given below :

- — Analysis of current energy consumption pattern and its past trends in details.
- — Review of heating and lighting requirements for an industry.
- — Review of existing energy recording systems of an industry.
- — Review the records of preventive maintenance of an industry.
- — Review of fossil fuels storage and handling adopted in an industry.
- — Review the new projects with respect to energy end-uses.
- — Calibration of sub-meter used in an industry.
- — Comparison of standard energy consumption with actual consumption of an industry.
- — Comparison of energy consumption with other locations, other firms, previous periods and budgets.
- — Comparison of meter-reading against log books of an industry.

— Drawing the diagram of heat balance for the firm.
— Check the records against invoices.
— Check the capacity and level of efficiency of equipments.
— Check the working of automatic controls system.
– Check the frequency of energy reporting system.
— Examine the level of training for energy management staff.
— Examine and monitor new energy saving techniques.
— Examine the need for energy saving techniques.
— Determine adequacy of maintenance.
— Assessment of the need for improved energy instrumentation.
— Introduction of monitoring procedures for energy use/consumption.
— Introduction of life cycle costing for energy.
— Consider the changing MIS to include energy parameter.
— Develop energy use indices to compare performance/productivity.
— Consider the publicity compaigns of incentives and tax-exemptionn for pollution free machinery.
— Replacement over-rated capacity machines etc.

The Need for Energy Audit

In India, energy audit becomes an effective tool and unavoidable for all sections in view of energy crisis and hikes in the cost of different forms of energy. The needs of energy audit are as follows:

— to understand more about the different ways of thermal energy and primary fuels used in an industry.
— to identify the thrust areas where wastage of energy can occur and where scope for improvement of energy saving exists in any area.
— to keep alive variations which occur in the energy costs availability and reliability of sustainable energy.
— to decide an appropriate energy-mix in any process/method, and,
— to identify appropriate energy conservation techniques.

Type of Energy Audit

Energy audit can be classified from different angle for different purposes. They are as follows:

— preliminary audit,
— detailed audit,
— financial energy audit,
— managerial energy audit etc.

Of all the types, the preliminary audit and detailed audit which are discussed below, are most significant and popularly used in many industries.

The preliminary energy audit is to make a quick estimate of energy requirements and energy efficiency of the process in a limited period of time. This is performed over a short span of time to focus major energy supplies and demands. It usually covers major equipments which account for nearly about 60-70% of the total energy consumed. The data needed are usually taken from production records. It is one of the initial steps taken on energy conservation programme. The frequency of preliminary audit can not be decided and pre-audit visit is also not at all required. However, a detailed questionnaire is to be compiled before the audit. It may form the basis for deciding the modolities of detailed energy audit.

The detailed energy audit shall be performed in a systematic manner for maximum energy saving. It goes beyond quantitative estimates of costs and savings. It should cover more than 90% of the energy consumed. A long term plan can be drawn upon on the basis of data generated and analysed in the detailed audit. This includes engineering recommendations and well defined projects with priorities like costs and savings, use of alternative energy etc. For energy intensive industries, the frequency of this audit may be on an annual basis whereas for other industries, it may be once in two to three years. One or two pre-audit visits are required. In order to conduct detailed audit, a detailed questionnaire should be compiled and advance notice to all departmental heads are to be issued. The detailed energy audit helps in the following ways:

— to evolve norms for energy consumption.
— to establish the baseline for measuring performance.

— to estimate energy cost as a percentage of value added cost.
— to identify the areas which hold higher potential for energy saving.
— to assess economic evaluation of different energy consumption opportunities.
— to train plant personnel in energy conservation techniques.
— to study the role of computers in a energy audit.

CONCLUSION

Energy audit is an effective energy conservation tool to define and pursue a comprehensive energy management programme in an industry. A careful audit could effectively manage the energy system at minimum energy costs for monitoring the energy consumption. It is certain that energy audit would be useful for industries in combating esculating energy costs and also reap several other benefits like improved production, better quality, higher profits, lower emission etc. The energy audit would broadly be the same in any type of industry and service. The basic formats may have to be suitably modified for different types of industries. It is an urgent need for all industries to introduce energy audit without fail and Government of India by regulation should also insist upon the same as compulsory for all industries. Further annual energy audit has to be made mandatory for all industries to conserve the precious and non-replenishable resources of our nation.

REFERENCES

"An Approach to Energy Audit", Petroleum Conservation Research Association, New Delhi.

"Energy Audit Directory" Petroleum Conservation Research Association, New Delhi.

Energy Audit Report, Aug. 1989, BHEL, Hyderabad.

Training programme on Energy Audit and Conservation for IDBI TCO personnel. (Back ground material), Vol. 1, National Productivity Council, New Delhi.

"Active Conservation Techniques", Petroleum Conservation Research Association. New Delhi.

R. Kannan, Energy "Conservation through Energy Audit in Central Workshops, Goldern Rock", *M. Tech. Project,* (unpublished), Deptt. of Chemical Engineering, REC, Trichy, Dec. 1996.

Index